Hegel's Actuality Chapter of the *Science of Logic*

Hegel's Actuality Chapter of the *Science of Logic*

A Commentary

Nahum Brown

LEXINGTON BOOKS
Lanham • Boulder • New York • London

Published by Lexington Books
An imprint of The Rowman & Littlefield Publishing Group, Inc.
4501 Forbes Boulevard, Suite 200, Lanham, Maryland 20706
www.rowman.com

6 Tinworth Street, London SE11 5AL

British Library Cataloguing in Publication Information Available

Library of Congress Cataloging-in-Publication Data

Name: Brown, Nahum, author.
Title: Hegel's actuality chapter of the science of logic : a commentary / Nahum Brown.
Description: Lanham, MD : Lexington Books, 2018. | Includes bibliographical references and index.
Identifiers: LCCN 2018046750 (print) | LCCN 2018048342 (ebook) | ISBN 9781498560573 (electronic) | ISBN 9781498560566 (cloth : alk. paper)
Subjects: LCSH: Hegel, Georg Wilhelm Friedrich, 1770–1831. | Modality (Logic) | Hegel, Georg Wilhelm Friedrich, 1770-1831. Wissenschaft der Logik.
Classification: LCC B2948 (ebook) | LCC B2948 .B765 2018 (print) | DDC 160—dc23
LC record available at https://lccn.loc.gov/2018046750

∞™ The paper used in this publication meets the minimum requirements of American National Standard for Information Sciences Permanence of Paper for Printed Library Materials, ANSI/NISO Z39.48-1992.

Printed in the United States of America

Contents

Abbreviations

WORKS BY HEGEL

EL	*The Encyclopaedia Logic*. Translated by T. F. Geraets, W. A. Suchting, and H. S. Harris. Indianapolis: Hackett Publishing, 1991.
PS	*Phenomenology of Spirit*. Translated by A. V. Miller. Oxford: Oxford University Press, 1977.
SL	*Hegel's Science of Logic*. Translated by A. V. Miller. Amherst: Humanity Books, 1969.
WL	*Werke in zwanzig Bänden, 6: Wissenschaft der Logik II.* Frankfurt am Main: Suhrkamp Verlag, 1969.

COMMENTATORS OF THE ACTUALITY CHAPTER

SC	Burbidge, John W. *Hegel's Systematic Contingency*. New York: Palgrave Macmillan, 2007.
CC	Di Giovanni, George. "The Category of Contingency in the Hegelian Logic." In *Art and Logic in Hegel's Philosophy*. Edited by Warren E. Steinkraus, 179–200. Atlantic Highlands: Humanities Press, 1980.
NC	Houlgate, Stephen. "Necessity and Contingency in Hegel's *Science of Logic*." *Owl of Minerva* 27 (1995): 37–49.

FM	Lampert, Jay. "Hegel on Contingency, or, Fluidity and Multiplicity." *Bulletin of the Hegel Society of Great Britain* 51.2 (2005): 74–82.
CM	Longuenesse, Béatrice. *Hegel's Critique of Metaphysics*. Translated by Nicole J. Simek. Cambridge: Cambridge University Press, 2007.
HO	Marcuse, Herbert. *Hegel's Ontology and the Theory of Historicity*. Translated by Seyla Benhabib. Cambridge: MIT Press, 1987.
MN	Stekeler-Weithofer, Pirmin. "Hegel on Reality as a Modal Notion." In *Hegel's Analytic Pragmatism*. Unpublished. Accessed online: http://www.sozphil.uni-leipzig.de/cm/philosophie/files/2012/11/StekelerHegelsAnalyticPragmatism.pdf

OTHER WORKS

CWA	Aristotle, *The Complete Works of Aristotle*, 2 volumes. Edited by Jonathan Barnes. Princeton: Princeton University Press, 1984.
CPR	Kant, Immanuel, *Critique of Pure Reason*. Translated by Paul Guyer and Allen W. Wood. Cambridge: Cambridge University Press, 1998.

Acknowledgements

I would like to thank Jay Lampert, John Russon, John W. Burbidge, Edward S. Casey, and Kelly Coble for taking the time to comment on earlier drafts of this book. Their advice has been invaluable for this work. I would also like to acknowledge the *International Association for Philosophy and Literature* and the *Ontario Hegel Organization* for inviting me to present conference versions of earlier drafts.

Introduction

The "Actuality" chapter of Hegel's *Science of Logic* has received significant attention in recent years, primarily because one can find in it the claim that "contingency is necessary." One of the main reasons why scholars have been drawn to the chapter is because, if contingency is necessary to Hegel's system, the case can then be made that the concepts of the *Logic*, far from being permanently fixed by the design Hegel had initially proposed for them between 1812 and 1816, contain within them a powerful dynamism, alterability, and propulsion to-be-otherwise. There are those who believe that the system is mostly closed, that necessity dictates the organization of the concepts, and that, with the exception of minor revisions and points of clarification, the *Logic* is an already complete ontology of being and actuality. But then there are those who believe that what Hegel has discovered through the system of the *Logic* is a formulation of necessity that is at the same time open to further contingencies. From this standpoint, the *Logic* offers a system of thought that cannot be otherwise but that is equally propelled in this necessity toward contingency. In this sense, the *Logic* offers us an ontology of becoming and difference, as much as an ontology of being and actuality.

This book explores Hegel's theory of modality through extremely close textual analysis of the "Actuality" chapter. The "Actuality" chapter is the equivalence of Aristotle's momentous *Metaphysics* book *Theta* (Book 9). Because of this, the chapter deserves the same thorough investigation into its complex insights and argumentation. This book situates Hegel's modal ontology within historical and contemporary debates about metaphysics, while analyzing some of the most controversial themes of Hegel's theory, such as the question of the ontological status of unactualized possibilities, the relationship between contradiction and possibility, and the claim that necessity leads to freedom. This book also contributes to an ongoing philosophical inquiry into the nature of dialectics by articulating Hegel's "Actuality" chapter as a coherent argument divided into twenty-seven premises.

Divisions of This Book and How to Read It

This book has been divided in two ways. Each of the three chapters presents a different type of modality (formal, real, and absolute modality). Formal modality has to do with the formal law of non-contradiction. Real modality has to do with the potentiality and actuality of the real, contextual world when recognized from the terms of determinate content. And absolute modality has to do with the relationship between modality and substance. Although each type of modality can be conceived of as a separate theory of modal reality, each should also be recognized as part of the basis for one overarching argument. To emphasize the coherence of the argument as a whole, this book has been divided, not only into three chapters, but also into twenty-seven premises and six remarks, which range over all three chapters. Each premise functions like a sub-chapter, and each remark either situates Hegel in terms of another philosopher or explores a moment of Hegel's thought in more depth than the premise-structure could sustain without losing the thread of the argument. The layout of the remarks imitates, stylistically, Hegel's own use of remarks, which appear scattered throughout the *Logic* and offer pathways, clarifications, and detours into topics that exceed the official trajectory of the *Logic's* conceptual development. As for the layout of the premises, Hegel does not illustrate this division of his argument explicitly, so it is up to us to recognize the important transitions of his argument and to discover the dialectic as it unfolds. The close textual analysis I offer in the premises contributes to a growing body of literature about the logical nature of dialectics. By standardizing Hegel's argument, I have attempted to avoid preconceptions about dialectical thinking, whether this means forming a triad or finding a synthesis behind a thesis and an antithesis. I often attempt to anticipate the best version of Hegel's position by analyzing how his various insights could be true. I also attempt analysis as critique by looking for gaps in the argument, testing his claims, and exploring examples. Each premise begins with a passage from Hegel's "Actuality" chapter. The appendix at the back of the book is also there as a reference tool for readers who would like to visualize more of the complexity of Hegel's argument. I have attempted to discover the argument in the course of the dialectic; nevertheless, some of the transitions of Hegel's argument are quite difficult to explain, especially when Hegel turns to absolute modality. I also realize that readers might propose the argument should go somewhat differently, that one should either add, subtract, or otherwise replace one premise with another. This, I contend, is the interpretive nature of the analysis.

Certainly, some readers will be disinclined to think that Hegelian arguments can be presented in terms of premises. Some readers will feel that the premise-design of this work makes his thought appear to be too static and inorganic. I think that the premise-design is one of the original contributions of this book, and that since it is original, it is in a way contentious and will lead to disagreement and debate about whether Hegel's speculative arguments can be analyzed in this way. This, I maintain, is one of the reasons why this book is important to scholarship. The premises attempt to highlight that Hegel is involved with rational argumentation and that we can analyze his arguments step by step. The disadvantage of this approach is that some readers will conclude that Hegel's dialectical, fluid way of thinking requires non-linear, para-rational formats of analysis, and that rendering his argument into premises is an un-Hegelian way of doing analysis. The advantage of this approach, nevertheless, is that we become able to critically analyze dialectical argumentation in a lot more detail and with a lot more clarity. I do not think that the premises are static or un-Hegelian, as some readers might suggest. On the contrary, I think that the premise-design helps us to articulate the mechanics behind Hegel's methodology more generally. In the introduction to the *Logic*, Hegel defines the term "logic" as the co-generative unfolding of form and content. The operation of thinking and the content of thought are immanently intertwined in such a way that each generates the other by generating itself. This means that the *Logic* as a book advances with the articulation of its methodology. I maintain that the premise-design enhances this methodological strategy because we become able to expose the intense detail in the text that the *Logic* demands of us in all of its complexity and with a lot more vividness.

To give the reader a visual of the book's chapters, I offer the following summaries:

Chapter 1 explains Hegel's initial account of formal, or logical, possibility, anticipating how Hegel incorporates a traditional definition of possibility as whatever is non-contradictory into his own argument, but ultimately demonstrating why Hegel views this definition as an assumption that must be overcome. This chapter contains the first thirteen premises of Hegel's argument. Premises one through six outline the main problem of the chapter, that although actuality contains possibility within it, it would seem that it cannot contain the diversity of the possible without becoming contradictory. Premises seven through thirteen then establish Hegel's rejection of this position and offer an alternative, formal solution, which ultimately also fails to accurately describe why actuality and possibility are transitional concepts.

Chapter 2 contains premises fourteen through twenty-two of Hegel's argument, outlining his primary solution to the formal problem that there is no immediate way to actualize the diversity of the possible in one actuality. This begins from a modal interpretation of context-related possibilities. But the clue to Hegel's solution comes most prominently from his definition of conditions as both actuality and possibility together. A condition is an immediate actuality, but since it is also the possibility of other actuals, it is a gateway between actuality and possibility. This chapter analyzes Hegel's theory of dispersed actuality, where the possibilities of one thing are contained in the actualities of others, his theory of conditional actualization, where something initial becomes actual through a series of possibilities, as well as his theory of compulsive necessity, where possibilities are embedded in other actuals and must be drawn out, but gain from this otherwise inaccessible formations of determinate content.

Chapter 3 culminates in Hegel's definition of substance as the complete, explicit identity between actuality and possibility. Premises twenty-three through twenty-seven demonstrate why substance results from real modality in the first place, how to think of a many-substance system as embedded within a one-substance system, and why the absolute necessity of total inclusion invokes contingency as a further consequence. This chapter offers evidence for Hegel's two primary insights about modality. Because of substance, actuality and possibility are explicitly transitional concepts. Because of the dialectic between necessity and contingency, necessity reveals itself as freedom.

The "Actuality" Chapter in Context

The Greater Logic contains a daunting array of themes, with hundreds of complex arguments, written in what might seem to be an overly dense conceptual language. Hegel divides the book into two volumes: into an objective and subjective logic. But he also divides it into three doctrines: into the *Doctrine of Being*, the *Doctrine of Essence*, and the *Doctrine of the Concept*. Hegel further divides the *Doctrine of Being* (first published in 1812) into "Quality," "Quantity," and "Measure"; the *Doctrine of Essence* (first published in 1813) into "Essence as Reflection Within," "Appearance," and "Actuality"; and the *Doctrine of the Concept* (first published in 1816) into "Subjectivity," "Objectivity," and "the Idea."[1]

One common way to read the book is to view it as a full-scale deduction of the fundamental categories of thought and reality.[2] Much like Kant's *Critique of Pure Reason* (A edition, 1781; B edition, 1787), the *Logic* aims to expose the necessary conditions of possibility and to uncover the universal categories of reality. But the *Logic* presents these conditions in terms of being rather than in terms of experience. In this sense, Hegel's method is speculative rather than transcendental. By analyzing the conditions of experience, Kant discovers in the transcendental deduction of the *First Critique* twelve necessary categories for the possibility of any experience whatsoever. This makes his method transcendental. However, by analyzing the basic conditions of being, Hegel reveals hundreds of categories (I call these concepts in Hegel) through his exploration of the nature of thought and reality.[3] The *Logic* can be viewed, in this respect, as one of the most comprehensive books of ontology because it aims to begin a deduction of the concepts of being from being alone. This makes the method speculative, exposing a self-referential treatment of logic that takes the negative into account rather than the purely positive identity of propositions and things. Since his analysis of being is immanent to being, each concept already contains the material for other concepts as part of what each is. The method is then analytic in the sense that each concept already implicitly contains what follows from one relationship to another, but this also makes the method synthetic in the sense that the reality of what is becomes determinate only by reckoning with the negative.

Hegel proposes in the *Logic* to begin from a rigorous, presuppositionless starting point in being, nothing, and becoming.[4] Hegel challenges us as readers to have the resolve to let go of all assumptions and opinions of what we think we know, initiating a kind of investigation of first philosophy similar to that of Socratic method or Cartesian doubt. The project that Hegel has designed of a science that begins without presuppositions can be conceived of as so radical that it does not even presuppose the most basic of axioms but must let these appear, if they ever do appear, in the course of thought's own systematic development. It would be a presupposition, for example, to list some quality of determinate being, such as Thales's theory of water, or to begin from a series of common sense definitions, which let other disciplines come to practical conclusions about the world, such as the species-genus distinction in biology or the basic laws of physics. For Hegel, even Descartes's *cogito ergo sum* tacitly posits assumptions about self-consciousness and duality.[5] If the activity of thought is to expose the reflection of its own content in a rigorous way, and therefore establish the forms of thought required of logic, we cannot even propose a list of categories as Kant does in his "Table of Categories" in the "Transcendental Analytic."[6] We must be able instead to deduce these from the spontaneous act of thinking. Without listing the forms of thought from the outset, we must let them appear in an immanent way and develop from a content that is at the same time their form.[7]

The relationship between Hegel's *Phenomenology of Spirit* (1807) and the *Science of Logic* complicates Hegel's assertion that the opening of the *Logic*—which transitions from being (*Sein*) to nothing (*Nichts*), and from nothing to becoming (*Werden*)—is the true starting point of his system. Hegel explains in the "Introduction" to the *Logic* that the *Phenomenology* should be viewed as a necessary preparatory work that clears away the assumptions of subject-object dualism and thereby clears away our deeply-sedimented prejudices about the inherent and seemingly insurmountable difference between consciousness and the world, a prejudice which was fully articulated by Descartes in the sixth meditation of the *Meditations* as the dualism of thought and extension. The primary consequence of the *Phenomenology*, in this regard, is to release us from the biases of everyday experience and to prepare us to be able to grasp the pure, logical standpoint of thought thinking itself and reality realizing itself from the rigor of a presuppositionless science. Hegel clarifies that although this means that the *Phenomenology* does, in one sense, precede the *Logic,* and that the *Logic* does, in a way, presuppose the grand narrative of the *Phenomenology*, the activity of clearing away subject-object dualism, which is successful, according to Hegel, with the final conclusion of "Absolute Knowing," should not distract us from the ultimate claim that the *Logic* begins without any presupposition whatsoever. As with all of the concepts that unfold presuppositionlessly from the opening of the *Logic*, Hegel's unusual analysis of the modal concepts is shaped by his radical proposal of a starting point that makes no assumptions.

The concepts of the *Logic* range from the simple structures of immediacy, the initial determinations of being, of quality, limit, and finitude, as well as the basic determinations of quantity and measure such as indifference, standing-beside-one-another, number, and arithmetic, to the more complex determinations of essence, where a thing has properties and can maintain a consistent identity throughout various fluctuations; relationships of reflection, identity, difference, contradiction, appearance, existence, part and whole, possibility and actuality, cause and effect, and reciprocity. We can view the *Doctrine of Being* as an exposition of the most basic concepts of being, as what is generated from an investigation into *that which* being fundamentally is. We can view the *Doctrine of Essence*, in contrast, as an exposition of the more complex concepts that are produced from the question of *what* being is, rather than from the surface-level description of *that which* being is. Essence emerges as the substratum of being, both as what being fundamentally is in essence, but also as the alienation of being from itself. We can then view the *Doctrine of the Concept* as an exposition of those concepts that arise from the reconciliation of being and essence together, as the resolution of this alienation, but also as the recognition that the essence of being is nothing other than being's own self-revelation.

The chapter "Actuality" appears as the penultimate chapter of the *Doctrine of Essence* under the section with the same name, "Actuality." It comes after the "Concrete Existence" and "Appearance" chapters, before the "Substantiality," "Causality," and "Reciprocity" subchapters of "the Absolute Relation," and is situated alongside a revealing remark on Spinoza and Leibniz. "Actuality" belongs to the *Doctrine of Essence* because what is actual has emerged from a prior source in possibility. This source is the modality of essence. What *is* as simple immediate being has nevertheless emerged into actuality from the negativity of the possible. Although for this reason the

"Actuality" chapter belongs to the *Doctrine of Essence*, this chapter is also an intermediary chapter between essence and the concept. Because actuality maintains the contraries of itself in possibility, it is both itself and the opposite of itself, but it is this *as itself*, that is, as the concept.

In Defense of Local Arguments

Because of the book's complexity, and because Hegel presents the *Logic* as one complete, systematic, and developmental account of thought and reality, with the implication that not any one division of the book should take precedence over any other, commentators of Hegel have largely objected to the strategy of isolating any specific argument or theme of the *Logic*, proposing instead that interpretations should appeal to the broadest consequences of Hegelian thinking generally. To isolate one chapter of the *Logic*, in this case "Actuality," inevitably raises questions about whether Hegel's arguments can be discussed out of the context of the book's overarching trajectory. Commentators who have this objection argue that isolating Hegel's arguments threatens to dislocate the themes of the book from the book's systematic purpose. The prevalence of this objection has left many of Hegel's local arguments under-analyzed.

Stephen Houlgate articulates this objection quite well and also offers a strong defense against it in the "Introduction" to his book *The Opening of Hegel's Logic*. "It is clearly very tempting when approaching Hegel," Houlgate writes, "to think that the whole picture is actually of primary importance and that the details of individual arguments are secondary or even incidental—mere "moments" of a totality that constitutes the real truth or mere "examples" of some universal, omnipotent dialectical principle."[8] Houlgate's primary defense against this criticism, which he uses to establish his project of analyzing the opening movements of the *Logic*, is to point out that there is a double-standard:

> When we read Descartes' *Meditations* and Spinoza's *Ethics*, we are urged to weigh individual arguments very carefully and consider whether or not they are valid. But when it comes to Hegel, the main point in the eyes of many seems to be to get a rough sense of the whole forest and not to worry too much about the trustworthiness of the individual trees. . . . If we are to follow Hegel himself [however] . . . the properly philosophical way to approach his texts is not to look in the prefaces and introductions for intimations of his general conception of dialectic or spirit, but to look in the main body of his texts at the many particular analyses.[9]

While there are certainly disadvantages to isolating one chapter of the *Logic*, I generally agree with Houlgate that scholarship on Hegel also needs to be able to analyze local arguments, and that we lose track of some of the most important insights of Hegel's thought if we limit ourselves to only the broadest statements about Hegelian philosophy. The modal argument in the "Actuality" chapter is one such local argument from Hegel that needs to be treated in detail. After all, the "Actuality" chapter is Hegel's version of Aristotle's *Metaphysics* book *Theta*. Hegel's argument should be scrutinized and dissected with the same care that centuries of commentators from Aquinas[10] to Heidegger[11] have given to Aristotle's stand-alone arguments from modality. What this book offers, then, is an extensive line-by-line commentary of Hegel's modal argument, with the aim of contributing to a growing body of textual and critical analysis about the "Actuality" chapter, which can be situated from within the complex mechanics of the *Logic* and of Hegelian thinking as a whole, but can also be treated as a stand-alone argument.

In summary, because this book focuses primarily on one theme, modality, readers might worry that the book will not be able to establish the kind of global perspective that the *Logic* requires. Hegel is, in this sense, a different kind of thinker from most other Western philosophers. But I contend that this is the sort of issue that everyone who works on a theme within the *Logic* has to face. The *Logic* is a complicated, powerful, endless book. One can attempt to approach the book globally, but one can also attempt to approach the book thematically. I have chosen to do the later. While I recognize that establishing global interpretations of Hegel's thinking is generally important for scholarship, I also think that it would be problematic to limit Hegel scholarship to only the kind of commentary that addresses the book as a whole. Hegel scholarship needs also to be able to allow for book-length studies of specialized themes within the *Logic*. By saying this, I do not mean to suggest that scholarship should be able to treat isolated chapters of the *Logic* without regard for the system that unfolds with the *Logic*, or without a sensitivity to what role such a theme plays in Hegel's corpus. But I think that over-contextualization, where themes are treated always only in terms of the *Logic* in its entirety, threatens to lose track of some of the most detailed, far-reaching consequences of Hegel's thought.

An Overview of Commentators

There are by now a number of excellent commentaries that each in their own way address Hegel's "Actuality" chapter. Herbert Marcuse's 1934 *Hegel's Ontology and the Theory of History* and Dieter Henrich's 1971 "*Hegel's Theorie über den Zufall*" in *Hegel im Kontext* are both pioneering studies that have prompted further lines of investigation. The analyses of Hegel's argument that I have found most compelling include Jay Lampert's 2005 essay "Hegel on Contingency, or, Fluidity And Multiplicity," John W. Burbidge's 2007 book chapter "The Necessity of Contingency" in *Hegel's Systematic Contingency*, Stephen Houlgate's 1995 essay "Necessity and Contingency in

Hegel's *Science of Logic*," George di Giovanni's 1980 essay "The Category of Contingency in the Hegelian Logic," Pirmin Stekeler-Weithofer's chapter "Hegel on Reality as a Modal Notion" in *Hegel's Analytic Pragmatism*, and Béatrice Longuenesse's book *Hegel's Critique of Metaphysics*, originally published in 1981 in French as *Hegel et la Critique de la Métaphysique*. I will briefly discuss each of these.[12]

My initial motivation in writing this analysis of the "Actuality" chapter comes from Lampert's essay on contingency. Lampert claims that because of the necessity of contingency, the truth of each thing is in the possibilities of others, which, when recognized from the disposition of the absolute, presents us with a system of free movement. For Lampert, the "Actuality" chapter is about how the "external differences" that seem to exist between things become a complex network of "internal multiplicity" and necessary propulsion (FM 75). Lampert begins from the theory that things are only possible if they can express the totality of the absolute from their own perspectives. He concludes from this that because each thing is not only itself, but also the expression of the whole, everything must interact in every possible way. Lampert claims that his reading would probably not have been possible prior to Deleuze.

Burbidge's book *Hegel's Systematic Contingency* is a major contribution to the literature on Hegel's "Actuality" chapter. In "The Necessity of Contingency" chapter, Burbidge claims that the conclusion from Hegel's modal argument is that "contingency is absolutely necessary." He sees the three moments of the thesis that contingency is necessary as the primary dialectical motor that motivates the transition points of Hegel's argument—first, the formal version of the thesis, that whatever happens to be actual (contingently) is nevertheless necessary simply because actuality has always already happened and cannot be undone—second, the real version, that relative necessity requires contingency as its starting point—and third, the absolute version, which states that the whole picture requires contingency as the basic feature for the recognition of a disposition of totality. Burbidge also makes a second outstanding contribution when he emphasizes why Hegel consistently begins from actuality. Burbidge claims that one can only give an account of possibilities after one gives an account of actuality. This is especially important for surveying what is really possible from what is actually present in reality.

Houlgate's equally excellent essay, "Necessity and Contingency," purports to clear up a misconception about Hegelian necessity. Houlgate rejects the notion that Hegel's theory of necessity makes him a determinist, claiming instead that there is "no independent power of necessity in Hegel's universe which determines all that occurs" (NC 45). Houlgate's analysis of the relationship between systematization and contingency is original because he believes that absolute necessity generates its own special type of content without, however, dictating what can or cannot come into existence (NC 45). In this sense, Hegel offers a compelling answer to the classic question of metaphysics, why is there being instead of nothing? Because of absolute necessity, being prevails over nothing. Houlgate sees this as the primary stance of necessity, that being itself cannot be contingent (NC 46). I am especially sympathetic of Houlgate's reading of Hegel's passages about absolute necessity, that all finite things necessarily perish. The inevitability of the destruction of determinateness is at the same time the expression of contingency, that things cannot do otherwise than to be otherwise (NC 47). Ultimately, Houlgate finds in absolute necessity not the restrictive blind necessity that we commonly conceive of as the antithesis of spontaneous freedom. He finds, instead, a theory of necessity that is the foundation of freedom. Because things cannot be otherwise than being, this lets them go free as being.

In "The Category of Contingency in the Hegelian Logic," Di Giovanni focuses on the contingency passages of Hegel's argument. He begins from a definition of contingency as an object that is ambiguously both actuality and possibility. Contingency is the disposition of an object that is "neither quite actual yet, nor possible" (CC 186). This treatment of contingency relies heavily on a Kantian reading of Hegel's argument. Di Giovanni's reading is Kantian in two senses: (1) He interprets Hegel as following Kant's method of taking an object and then ascertaining whether it is actual or possible (most other commentators of Hegel's modal argument do not follow this method, at least not explicitly). (2) He sees Hegel's modal theory as offering an account of the "reflective awareness of the nature and the limits of the experience" (CC 194). Although I find (1) problematic because it obscures the discovery of certain important premises of Hegel's argument, especially where Hegel says that actuality and possibility transition into each other, Di Giovanni's final conclusion about the role of contingency is quite insightful. Against what he posits as the "classical metaphysical" position, which views contingencies as disturbances or digressions of reality (CC 186), Di Giovanni proposes that for Hegel contingencies belong to the property of reality, no matter how unpredictable or inconsistent they may appear to be. This can be seen as Di Giovanni's interpretation of the thesis that contingency is necessary.

One of Stekeler-Weithofer's main insights in his commentary of the "Actuality" chapter in *Hegel's Analytic Pragmatism* comes from situating Hegel's modal theory in terms of formal modal logic. He says that the modal logicians follow the Tractarian concept of possibility as "truth functionally composed on the basis of logically elementary propositions" (MN 219). But Stekeler-Weithofer believes that Hegel's project is about modal reality, how things, rather than propositions, are constituted in terms of modality. We can think of this distinction from Stekeler-Weithofer as the difference between *de dicto* and *de re* consequences (broadly construed). An analysis of possibility and necessity from the terms of *de dicto* consequences gives access to certain projects that concern logic and language. But if we think of possibility and necessity from the terms of *de re* consequences, this leads to modal insights about the nature of reality, such as how to conceive of something as both an instance of essence and the essence itself.

Béatrice Longuenesse has a particular way of dividing the formal, real, and absolute sub-chapters of Hegel's text. She claims in "What Is Rational Is Actual" of *Hegel's Critique of Metaphysics* that with formal modality, Hegel rejects the classical axioms and definitions of modality, which she says come mainly from Aristotle. With real modality, Hegel then analyzes Kant's modal theory. And finally with absolute modality, he develops his own theory (CM 119). While this is certainly a plausible interpretation of Hegel's design for the division of the sub-chapters, I think this division also presents the reader with an opposition between the sub-chapters which I doubt Hegel had fully intended. In contrast to Longuenesse, I treat the "Actuality" chapter as one argument, where the premises and conclusions of the formal and the real sub-chapters eventually give way to the premises of absolute modality. Longuenesse's analysis is especially adept at integrating and contrasting Hegel's theory with Spinoza (CM 153–58), Leibniz (CM 132–33), and Kant (she sustains a contrast of Kant with Hegel throughout her entire chapter). By interpreting Hegel's absolute necessity as the relationship between being and thought, Longuenesse effectively demonstrates the transition that Hegel discovers from the *Doctrine of Essence* to the *Subjective Concept*. While Lampert has emphasized "multiplicity," Houlgate has emphasized "absolute necessity," and Burbidge and Di Giovanni have each emphasized "contingency," Longuenesse sees the development of Hegel's argument as based on how to give a complete account of "actuality." In contrast, my thesis is that the motivation of Hegel's argument comes from the relationship between modality and contradiction, where actuality is continuously reformulated to include contrary possibilities as part of the constitution of modal reality.

Hegel's Argument from Modality (The Short Version)

Hegel's argument begins from two self-evident premises. On the one hand, what is actual is existence (premise one). On the other hand, what is actual is possible (premise two). Although each premise is self-evident, when thought attempts to think both premises together, a complication arises. If what is actual is existence, and yet what is actual is possible, then it would seem that the possible exists in the same way that the actual exists. This is problematic, however, because what is merely possible (for example, unicorns, worlds in which gravity goes up, science-fiction stories, and so on) does not seem to have the same ontological status as concrete actuality does. But since Hegel is committed to the reconciliation of both premises, he presents an argument for why we look to existence for the nature of possibility just as we look to existence for the nature of actuality. This calls for a revision of what it means to actualize possibility. Possibility and actuality become transitional concepts.

Hegel acknowledges that in an immediate sense it would be contradictory, and therefore impossible, to actualize the totality of possibility in one actuality (premises three to seven). If something is possible, it can and can not be; however, there is no way to actualize this contrariety without falling into contradiction. To overcome this problem, Hegel first proposes reflected actuality (premise eight), and then formal contingency (premise ten), as two different solutions for how to posit the existence of possibility in actuality without removing the contrariety of the possible. But reflected actuality does not work because it can only express the existence of mere unactualized possibility (premise nine). Contingency does a better job of integrating the possible in the actual because it holds together the contrary sides of possibility through indifference, in the sense that a contingent actuality posits other unactualized possibilities that could have been (premises eleven and twelve). However, since contingency posits these unactualized possibilities only through indifference, Hegel then turns to formal necessity as the complete coincidence of the actual and the possible together (premise thirteen). On the one hand, this necessity is like the principle of non-contradiction (from premise seven), in the sense that it restricts the possible to only what is actually possible. However, on the other hand, this necessity is the only form that actualization can take up to actualize possibility itself.

Hegel then turns from formal modality to real modality because there is no way from the formal structure to actualize across the totality of possibility. Real modality can more appropriately express the negativity of the possible because, by actualizing across content and not only across form, immediate actualities find their possibilities dispersed in others, and then literally find themselves in these others even though they are not these others. Hegel posits real actuality (premise fourteen) and real possibility (premise fifteen) in a relationship of existing multiplicity (premise sixteen), which leads in turn to his argument that each contextual thing has its possibilities dispersed in the actualities of others (premise seventeen). The advantage of dispersed actuality over the earlier formal solutions comes from the recognition that immediate, contingent actualities are the conditions for the further actualizations of others (premise eighteen). This is an advantage because a condition is both actuality and possibility together. However, since conditions begin as actualities with the possibilities of themselves as others embedded in their content, Hegel claims that real necessity is required to draw the possibility out of them (premise nineteen to twenty-one).

Once thought recognizes that real necessity is relative necessity, and that this process of actualizing across conditions begins from contingency (premise twenty-two), Hegel proposes a third version of modality, "absolute modality." Absolute modality is an advancement upon real modality because it can include the contrariety of the possible as constitutive of reality. This version of modality begins from an initial description of many substances as absolute actuality (premise twenty-three). What is absolutely actual includes all permutations of possibility as part of what the actual is. But since this actuality already contains all of its possibilities, it has necessity, not possibility, over against it, and can no longer be otherwise (premise twenty-four). However, the reason why absolute actuality can no longer be otherwise is because it includes possibility rather than restricts it. Hegel calls this disposition of inclusion "absolute possibility" (premise twenty-five). By introducing absolute possibility, thought comes to recognize that what had seemed to be the many substances of absolute actuality is really the one substance of absolute necessity (premise twenty-six). Because absolute necessity is the inclusive necessity of absolute possibility, it is as much contingency as it is necessity (premise twenty-seven). This leads to Hegel's conclusion that necessity is in a sense freedom because it generates otherwise inaccessible possibilities.

The thesis that actuality and possibility are transitional concepts can, therefore, be viewed as the motor behind Hegel's entire argument, in the sense that it is this thesis that generates all of the major transition points of Hegel's argument. That actuality cannot maintain the possibility to-be and the possibility not-to-be as one unity drives Hegel from an initial conception of immediate actuality to a contingent version of actuality, where the actual is no longer only one of many possibilities, but has become an actuality that equally posits the existence of its possible contrary. This same impasse also leads to the transition from formal to real modality, because only from the context of things in their inter-relations with others are we able to anticipate how something can be both itself and the opposite of itself without falling into contradiction. I emphasize that conditions are precisely this transitional movement between actuality and possibility, where each concept passes over into the other. Taken alone and in isolation, an acorn contains only the immediate form of its identity with-itself. But when we see the acorn as a condition in the context of others, we see it not only as something that is self-coherent, but also as the possibility of other actuals, as the possibility of a tree that could grow from it, or of a squirrel who could eat it. The thesis that actuality and possibility pass over into each other also generates the absolute modality premises of Hegel's argument. Actuality only becomes absolute when it can explicitly include the totality of possibility without falling into contradiction. This is the beginning of Hegel's theory of substance, of an actuality that is at the same time all permutations of itself in possibility.

NOTES

1. This discussion of the context of Hegel's "Actuality" chapter draws on Nahum Brown, "Indeterminacy, Modality, Dialectics: Hegel on the Possibility *Not to Be*," In *The Significance of Indeterminacy*, ed. Robert H. Scott and Gregory S. Moss (New York: Routledge, 2018), 104–23.

2. For example, see Stephen Houlgate, *The Opening of Hegel's Logic: From Being to Infinity* (West Lafayette: Purdue University Press, 2006).

3. Referring to the structure of the *Logic* in terms of the necessary conditions for the possibility of being inevitably invokes non-metaphysical readings of Hegel. Non-metaphysical readings, such as Robert B. Pippin, *Hegel's Idealism: The Satisfactions of Self-Consciousness* (Cambridge: Cambridge University Press, 1989) and Terry Pinkard, *Hegel's Phenomenology: The Sociality of Reason* (Cambridge: Cambridge University Press, 1994), offer a conceptual vision that situates Hegel's work within Kant's transcendental rejection of metaphysics in the *Critique of Pure Reason*. But I maintain that the reading of the "Actuality" chapter that my book promotes can also support metaphysical readings of the *Logic*. Metaphysical readings, such as Charles Taylor, *Hegel* (New York: Cambridge University Press, 1975), and Frederick C. Beiser, *Hegel* (London: Routledge, 2005), can be interpreted to view the large-scale developments of the *Logic* as the exposition of God's thought and of the true nature of reality.

4. Significant debates have surfaced about whether Hegel's starting point in the *Logic* is really presuppositionless, or whether Hegel relies on certain tacit assumptions without being aware of them. Four of these debates include (1) Does Hegel presuppose the disposition of absolute knowing that concludes the *Phenomenology of Spirit*? (2) Does Hegel presuppose dialectics from the outset? (3) Does Hegel's conception of being contain assumptions about universality, indefiniteness, and self-evidence, as Heidegger claims it does in the introduction to Martin Heidegger, *Being and Time*, translated by John Macquarrie and Edward Robinson (New York: Harper & Row, 1962), 22–4. (4) Does Hegel succumb to foundationalism, the criticism that any theory of first philosophy presupposes that there must be a starting point from which all things necessarily follow? Many of these debates are outlined in Richard Dien Winfield, *Hegel's Science of Logic: A Critical Rethinking in Thirty Lectures* (Lanham: Rowman & Littlefield, 2012), 17–45.

5. For his discussion of Descartes, see G. W. F. Hegel, *Lectures on the History of Philosophy: Volume 3, Medieval and Modern Philosophy*, translated by E. S. Haldane and Frances H. Simson (Lincoln: University of Nebraska Press, 1995).

6. Hegel claims that Kant "adopts [the categories] empirically, just in the way they have been ordered in logic." Hegel, *Lectures on the History of Philosophy: Volume 3*, 176.

7. For two excellent treatments of Hegel's presuppositionless starting point, see Houlgate, *The Opening of Hegel's Logic* and Winfield, *Hegel's Science of Logic*.

8. Houlgate, *The Opening of Hegel's Logic*, 4.

9. Houlgate, *The Opening of Hegel's Logic*, 5.

10. Saint Thomas Aquinas, *Commentary on the Metaphysics of Aristotle, Volume 1*, translated by John Patrick Rowan (Washington: H. Regnery Company, 1961).

11. Martin Heidegger, *Aristotle's Metaphysics Θ 1–3: On the Essence and Actuality of Force*, translated by Walter Brogan and Peter Warnek (Indianapolis: Indiana University Press, 1995).

12. Other works that I have found helpful include Gabriella Baptist, "Ways and Loci of Modality: The Chapter 'Actuality' in the *Science of Logic* between its Absence in Jena and its Disappearance in Berlin," in *Essays on Hegel's Logic*, ed. George di Giovanni (Albany: State University of New York Press, 1990), 127–44; Errol E. Harris, *An Interpretation of the Logic of Hegel* (Lanham: University Press of America, 1983); John F. Hoffmeyer, *The Advent of Freedom: The Presence of the Future in Hegel's Logic* (Cranbury: Associated University Presses, 1994); Iain Macdonald, "Adorno's Modal Utopianism: Possibility and Actuality in Adorno and Hegel," *Adorno Studies* 1 (2017): 2–12; John and Ellis McTaggart, *A Commentary on Hegel's Logic* (New York: Russell & Russell, 1910); G. R. G. Mure, *A Study of Hegel's Logic* (Oxford: Oxford University Press, 1950); Karen Ng, "Hegel's Logic of Actuality," *Review of Metaphysics* 63 (2009): 139–72; Taylor, *Hegel*; Tomoyuki Yamane, *Wirklichkeit: Interpretation eines*

Kapitals aus Hegels "Wissenschaft der Logik" (Frankfurt am Main: Peter Lang, 1983); and Christopher Yeomans, *Freedom and Reflection: Hegel and the Logic of Agency* (New York: Oxford University Press, 2012).

Chapter One

Formal Modality

1. What is actual is existence.

> Die Wirklichkeit ist formell, insofern sie als erste Wirklichkeit nur unmittelbare, unreflectirte Wirklichkeit, somit nur in dieser Formbestimmung, aber nicht als Totalität der Form ist. Sie ist so weiter nichts als ein Seyn oder Existenz überhaupt. (WL 202)

> Actuality is formal insofar as, being the first actuality, it is only immediate, unreflected actuality, and, therefore, is only in this formal determination but not yet as the totality of form. As such it is nothing more than a being or existence in general. [1]

Hegel's theory of modality begins from the premise that what is actual is existence in general. This premise can be interpreted in one of two ways. If Hegel is making a claim about actuality, the point is that we should look to existence to recognize what is actual. However, if he is making a claim about existence, then the point is that what exists has the character of being actual. I retain the first interpretation as the premise heading. That we should look to existence to see what is actual is an important claim that Hegel will develop over the course of his argument. This first interpretation of the claim helps us to answer the question, what is actuality? Actuality is existence.

To understand this, we will need to establish more about what Hegel means by existence (*Existenz*). In the "Existence" chapter (BK 2, SN 2, CH 1), which precedes the "Actuality" chapter, Hegel refers to existence both in its specificity, as the thing-emerging-process from which something comes into existence when it completes the totality of its conditions, but also in its generality, as the fact of existence. Premise one relies on both of these meanings of existence. Actuality is existence when it emerges from essence;

however, actuality is also primary, immediate, and unreflected because it is "mere existence," the immediacy of what is "simply there." We think of existence as what emerges into being from some prior source. We also think of existence as what needs no further explanation because it is already simply given.

We might assume that because existence and actuality are part of the *Doctrine of Essence* and not of the *Doctrine of Being*, existence should only refer to essence, to what things really are, to what emerges out of this immediacy, to *Existenz*, literally to "ex-istere," but not also refer to what is "simply there," to *Dasein*, literally to "being-there." It might seem inconsistent that Hegel initially defines actuality as immediate and unreflected (*unmittelbare und unreflectirte*). However, as we will see in the course of his argument, this is not merely an inconsistency in Hegel's writing. Hegel begins from an immediate, unreflected actuality because that which emerges is nothing other than the immediacy from which the actuality began.[2]

Formal actuality is the modal version of pure indeterminate being, the presuppositionless starting point at the very beginning of the *Logic*. The modal version of *to be* is *to be actual*. This initial actuality refers to the fact of existence as what is merely given without offering any further reason for why it is. This is why Hegel says that "[actuality] is nothing more than a being or existence in general" (WL 202). If I look out at the landscape, I see the trees, the hill, and the river as simply there, as part of the landscape, as a fact of existence, without their offering further recourse for why these features are what they are. But whereas existence is distinct from pure indeterminate being in the sense that its correlate is essence, actuality is distinct in the further sense that its correlate is possibility.[3] In other words, the concept of actuality is the same as the concepts of being and existence, with the one exception that while indeterminate being implies nothing (*Nichts*) and existence implies essence (*Wesen*), actuality contains the further implication that it has possibility (*Möglichkeit*) over against it. While the trees, the hill, and the river are merely given as part of the landscape, these features have equally emerged into actuality from possibility.

2. What is actual is possible.[4]

> Aber weil sie wesentlich nicht blosse unmittelbare Existenz, sondern, als Formeinheit des Ansichseyns oder der Innerlichkeit, und der Aeusserlichkeit ist, so enthält sie unmittelbar das Ansichseyn oder die Möglichkeit. Was wirklich ist, ist möglich. (WL 202)

> However, because by essence it is not merely immediate existence, but is rather the formal unity of the in-itself, or inwardness and outwardness, formal actuality immediately contains the in-itself or possibility. What is actual is possible.

Since actuality is immediate existence and also existence as emerging in essence, actuality contains possibility. Actuality and possibility are not separate concepts in their own right, but rather aspects of each other. This is important to Hegel's argument, as we will see at premises six and seven, because of the contradiction that emerges from what is initially an unsustainable identity between actuality and possibility. Rather than forming a distinct concept, possibility is the implicit in-itself of actuality, and actuality is the explicit externalization, or the outwardness, of what it already implicitly is.

However, Hegel is also saying that actuality is this whole process. The relationship between actuality and possibility is therefore already quite complicated. While actuality and possibility are not simply one, and in this way completely identical, they are also not obviously separable. This is apparent from Hegel's ambiguous terminology. First, he says "actuality contains possibility," but then he says "actuality is possibility." On the one hand, if actuality *contains* possibility, there is no position for possibility that is not already within actuality. This is why Hegel claims that possibility is the "in-itself" (*Ansichseyn*) of actuality. But on the other hand, if actuality is possibility, then possibility is the content of actuality, and actuality is only the manifestation of that which possibility already implicitly is.[5]

The last sentence of the passage is particularly significant. "What is actual is possible." We make a simple inference from the immediate, self-evidence of actuality, that whatever is actual must have been possible. If I am reading the newspaper, then reading the newspaper must have been possible and cannot have been impossible. I am reading it (this is actual), so I must be able to read, and the newspaper must be the sort of thing that can be read. Immediate actuality has this authority. Its possibility is obvious and cannot be contested. This is why we attribute definitive truth to the immediate presence of what is already there.[6] We can see in this way how premise two lends support to premise one. By recognizing that everything actual is possible, we affirm the obvious truth of premise one. Possibility is truth-affirming. Although we had already recognized actuality to be an existent, it is nevertheless the role of possibility to affirm this, that is, to offer evidence for actuality. In this way, Hegel anticipates the traditionally analytic definition of possibility as the minimum condition for the validity of a proposition.

But it is equally apparent that the possible is not necessarily actual. Actuality entails possibility, but possibility does not in the same manner entail actuality. This is one of the basic points Fitting and Mendelsohn make in their book *First-Order Modal Logic*. "Now, $p \supset \Diamond p$ (i.e., *It's actual, so it's possible*) is usually considered to be valid—Hughes and Cresswell (1968) call it the "Axiom of Possibility"—but its converse, $\Diamond p \supset p$ (i.e., It's possible, so it's actual) is not [valid]."[7] While the statement "it rains" necessitates "it possibly rains," the statement "it possibly rains" does not necessitate "it rains." This is the case not only for truth functional sentences but for entities and things in the world as well. Although I might buy a house next week, the possibility of this does not necessitate its actuality. This is more obvious when we think of examples of what is merely possible, where the possibility as such has never been demonstrated in actuality. Fantasy and science fiction stories offer countless examples of what is merely possible, of what has not yet and might not ever become actual. While we can speculate about these various possibilities, their mere projection in possibility does not necessitate their actuality.

Later, at premise ten, Hegel will argue that everything possible exists, and in the way that actuality exists, which will in effect reinstate the question of whether possibility entails actuality. But from Hegel's discussion at premise two, it is important to establish the obvious absurdity of the inference from possibility to actuality. The advancement of his argument, especially at premise eighteen, where a condition is a possible that entails further actuals, nevertheless depends upon the initial self-evidence that possibility does not entail actuality. The main force of development behind premises three to nine is to expose why thought must come to revise the relationship between actuality and possibility, to the point at which possibility does exist in the way that actuality exists.

3. If what is actual is possible, then the possible is the reflection of the actual into itself.

> Diese Möglichkeit ist die in sich reflectirte Wirklichkeit. Aber diß selbst erste Refectirtseyn ist ebenfalls das Formelle, und hiemit überhaupt nur die Bestimmung der Identität mit sich oder des Ansichseyns überhaupt. (WL 202–3)

> This possibility is actuality reflected into itself. But even this self-reflectedness is something formal and therefore only the determination, identity-with-self, or in-itself in general.

Hegel claims that the first function of possibility is to reflect actuality into itself. Since actuality is existence, possibility is only the affirmation of this existence after the fact. It is obvious that if I am actually reading the newspaper, this is possible. There seems to be a redundancy here. The possible does not add anything to the actual. It just affirms the identity of the actual, which is already there.

But this simple role as identity-with-self is more significant than one might think. If there were no such thing as reflected determinations, that is, if we were unable to infer possibility from actuality, we would think of actuality only in terms of the immediate surface, as in the *Doctrine of Being*, of the actuality that simply is. While this is right—actuality simply is—we recognize in the found authority of the actual, not only its surface immediacy, but equally its emergence into itself through the process of its actualization. Debates about whether possibility really exists arise precisely from this seemingly circular claim that actuality emerges into itself from the process of actualization. If actuality did not have possibility over against it as its reflected determination, we would be unable to recognize from this that actuality has emerged into itself from a prior source.

Because actuality entails possibility, possibility is the most basic condition for any actuality whatsoever. Possibility is in this sense a necessary requirement for actuality. Take the possibility away, and there goes the actuality as well. Just as actuality entails possibility (by *modus ponens*), in the sense that if we suppose something is actual, then it must be possible, likewise something that is not possible is not actual (by *modus tollens*), in the sense that if we suppose something is not possible, then it is not actual.

Still, possibility is not a sufficient requirement for actuality. If it were both necessary and sufficient, then possibility would entail actuality because the mere possibility of something would be sufficient enough to secure its actuality. This is why Hegel does not say that possibility is the identity of actuality, but only of actuality when it is reflected into itself (*in sich reflectirte Wirklichkeit*). Possibility is not the sort of other that stands against actuality and affirms its own identity as something that actuality is not. Rather, possibility is just the mirror-other of actuality, an other who is not an other, but only the reflection-into-self of actuality. From what would seem to be the redundancy of possibility's function for actuality, Hegel discovers an important transition. If possibility is the reflection of actuality into itself, this is because actuality must risk that it is not what it is. Possibility is then both the yes of affirmation, but equally the risk of the no. Actuality obviously entails possibility. But the non-existence—that this actuality could not have been—is equally entailed in the affirmation. Hegel discovers that by affirming the actual, the possible exposes the actual to the other of itself, and in this way serves to function as the maybe of the actual, both the yes and the no together.

4. But if the possible is only the reflection of the actual, it lacks itself and requires its completion in actuality.

> Die Möglichkeit enthält daher die zwey Momente; erstlich das positive, daß es ein Reflectirtseyn in sich selbst ist; aber indem es in der absoluten Form herabgesetzt ist zu einem Momente, so gilt das Reflectirtseyn-in-sich nicht mehr als Wesen, sondern hat zweytens die negative Bedeutung, daß die Möglichkeit ein mangelhaftes ist, auf ein anderes, die Wirklichkeit, hinweist, und an dieser sich ergänzt. (WL 203)

> Therefore, possibility contains two moments. First, it contains the positive moment of being the reflection of itself into itself. But because it has been

> reduced in this reflectedness to only a moment of the absolute form, it no longer has the status of essence, but has, as its second moment, the negative connotation of possibility as something incomplete, as something pointing to another, to actuality, to make it whole.

This first function of possibility, as the mere reflection of actuality into itself, leads to the further inference that what is merely possible lacks itself and requires its manifestation as actuality in order to complete itself. Premise four presents evidence for why the inference from actuality to possibility is not simply tautological. "What is actual is possible" is not tautological because the status of something possible is the status of something actual when it is not itself. Possibility serves a double function. On the one hand, it is the most basic condition for any actuality whatsoever. This is what Hegel calls the positive moment of possibility. But on the other hand, since possibility is only the in-itself of actuality, it is a reflection of actuality that is always incomplete, as the other of what is.

Hegel separates possibility into these two moments, but the separation is misleading. Possibility does not "contain two moments" in the way that we might think of a person's character as being in one way genuine, in another way disingenuous. The two moments arise, rather, from each other. Actuality and possibility come out of each other as co-dependent and co-generative aspects of the same concept.

What these positive and negative moments reveal is that possibility affirms the truth of actuality only if it equally affirms the negativity of the same actuality. In this sense possibility is both what the actual could be, but also the opposite of the actual, what it itself could not be. Again, if reading the newspaper is possible, this means both that I can actually read it, but also that I can *not* read it. The reason why Hegel then claims that possibility lacks itself and requires actuality to complete itself is because the co-positing of the affirming and the negating function places the possibility at a distance from its result in actuality.

5. Because it is reflection, possibility is the relating ground between the actual and the negation of the actual.[8]

There is a slight inconsistency in Hegel's argument at this point in the text. Hegel had called the positive and negative moments the moments *of possibility*, but if possibility is the comparing relation, then the positive and negative moments turn out instead to be the moments *of actuality*, and possibility turns out to be that which contains these moments.

> A ist möglich, heißt so viel als A ist A. Insofern man sich nicht auf die Entwiklung des Inhalts einläßt, so hat dieser die Form der Einfachheit; erst durch die Auflösung desselben in seine Bestimmungen kommt der Unters-

chied an ihm hervor. Indem man sich an jene einfache Form hält, so bleibt der Inhalt ein mit sich identisches und daher ein Mögliches. Es ist aber damit eben so Nichts gesagt, als mit dem formellen identischen Satze. (WL 203)

To say that *A is possible* is merely to say that *A is A*. If nothing is done to develop the content, possibility maintains the form of simplicity. Difference only emerges within it once it has been resolved into its determinations. As long as we hold to this simple form, the content remains something identical with itself and therefore something possible. But to say this is equally to say nothing, as with the formal law of identity.

Modality is formal if possibility is only about identity and does not change the content of something or introduce difference. The possible is the self-integrity of the actual. If the sea battle is possible, then it would not break the logical coherence of the event if it were to become actual. In this sense, the possible is already pre-disposed to become actual. There is no difference added to the content of the possible when it becomes actual. But to say that the sea battle is possible is only to say that if it were to become actual, this would not be impossible—which is to say nothing at all. To say "A is possible" is to express only the most empty of determinations, that what is possible can be, because if it were to become actual, this would be no different than this content as possibility.

Die Möglichkeit als aufgehoben gesetzte Formbestimmung, einen Inhalt überhaupt an ihr hat. Dieser ist als möglich ein Ansichseyn, das zugleich ein aufgehobenes oder ein Andersseyn ist. Weil er also nur ein möglicher ist, ist eben so sehr ein anderer und sein Gegentheil möglich. A ist A; eben so -A ist -A. Diese beyden Sätze drücken, jeder die Möglichkeit seiner Inhaltsbestimmung aus. Aber als diese identischen Sätze sind sie gleichgültig gegen einander; es ist mit dem einen nicht gesetzt, daß auch der andere hinzukomme. Die Möglichkeit ist die vergleichende Beziehung beyder; sie enthält es in ihrer Bestimmung, als eine Reflexion der Totalität, dass auch das Gegentheil möglich sey. Sie ist daher der beziehende Grund, daß darum, weil A= A, auch -A= -A ist; in dem möglichen A ist auch das Mögliche Nicht A enthalten, und diese Beziehung selbst ist es, welche beyde als mögliche bestimmt. (WL 204)

This finds expression at first in this way, that possibility as form determination posited as sublated possesses a content in general. As possible, this content is an in-itself which is at the same time something sublated or an other. But because this content is only a possible, another opposite to it is equally possible. *A is A*; equally, *-A is -A*. Each of these statements expresses the possibility of its content determination. But, as identical statements, each is indifferent to the other. That the other is also added is not posited in either. Possibility is the comparing relation of both statements; as a reflection of the totality, it implies by its definition that the opposite also is possible. Possibility is therefore the relating ground that, because A equals A, -A also equals -A; entailed

> in the possible A there is also the possible not -A, and it is this reference itself connecting them which determines both as possible.

Hegel had just explained possibility in terms of self-identity, that if A is possible, then A is A. But directly after this, he claims that if A is possible, it is not only self-identical, but the opposite of its identity is equally possible. As the most general projection of content, possibility is the relating ground between the two equally true yet empty articulations of the law of identity, that A= A but equally -A= -A. All possibilities of content are contained within these two articulations of the law of identity. The statement "if A is possible, then A is A" thus marks the initial transition from form to content. It starts the transition from formal modality to real modality.[9]

If A is possible, then -A is equally possible. Hegel claims from this that the possible A contains the possible -A within its own concept. The question, then, is whether the actuality of A, insofar as it is possibility, also contains in its own self-coherent identity the content of -A. Hegel's initial answer to this question is what establishes, as we will see, the contradiction that will arise at premise six.

Remark: Agamben on the Ability Not to Be

Hegel's definition of possibility as the relating ground between the actual and the negation of the actual can be viewed as a precursor to Agamben's work on potentiality, especially in terms of Agamben's analysis in *Nudities* of the ability *not to be*. In this short piece, "On What We Can Not Do," Agamben argues that the ability not to be is as constitutive of potentiality as the ability to be.[10] Similar to Hegel's claim that possibility is always both the affirmation and negation of the actual, Agamben claims that every potentiality (he uses the term potentiality rather than possibility) is also impotentiality. By impotentiality, he does not mean "not being able to" but rather "being able not to."[11] Agamben sees the can not as a potentiality in its own right. Just as someone might have the ability to build, this person can also exercise the ability not to build. Exercising the ability not to be undermines, for Agamben, the traditional authority of the actual.[12] We have the ability to withhold actuality. The strength of potentiality lies not only in the position of becoming actual, but equally in the position of not becoming actual.

This insight works against an assumption that we might otherwise have of Hegel's claim, that if something has the status of the possible, this status is only for the sake of the actual, as if the possibility not to be were really not a possibility at all, but only a deviation of the actual. Agamben thinks that human beings are different from animals precisely in this respect. Human beings can choose not to do or be. They can "see darkness."[13] "While fire can only burn," Agamben writes, "and other living beings are only capable of their own specific potentialities—they are capable of only this or that behavior inscribed into their biological vocation—human beings are the animals capable of their own impotentiality."[14]

Hegel's argument also works from the claim that what is possible enables the negation of actuality as certainly as it affirms the actual in existence. But Agamben can be interpreted to take this insight much further by exploring the productive political components of the potential not to be. The main insight from "On What We Can Not Do" comes in the form of a prognosis about the contemporary age. Although human beings are the animals who are capable of not acting, we have become alienated in the contemporary age from this capacity. We now act as if we can do anything. But this is precisely the loss of what makes us human, the potentiality to resist in actuality the process of actualizing this or that ability. Agamben prescribes that in this age we need to become sensitive again to our ability to resist actuality.[15]

6. If the possible A contains the possible -A, then what is possible is also impossible.

> So ist sie der verhältnißlose, unbestimmte Behälter für Alles überhaupt.—Im Sinne dieser formellen Möglichkeit ist alles möglich, was sich nicht widerspricht; das Reich der Möglichkeit ist daher die grenzenlose Mannichfaltigkeit. Aber jedes Mannichfaltige ist in sich und gegen anderes bestimmt und hat die Negation an ihm; überhaupt geht die gleichgültige Verschiedenheit in die Entgegensetzung über; die Entgegensetzung aber ist der Widerspruch. Daher ist Alles eben so sehr ein widersprechendes und daher unmögliches. (WL 203)

> Possibility is the relationless, indeterminate container for everything generally. In terms of formal possibility, everything is possible that does not contradict itself. The realm of possibility is a limitless multiplicity. But every multiplicity is determinate in itself and as against another and has negation in it. In general indifferent diversity passes over into opposition; but then opposition is contradiction. Therefore everything is just as much something contradictory and, because of this, impossible.

If something is possible, then it can be and also can not be. Therefore, everything is possible. When we hold the concept of possibility together as one unity, we transition between the particular level of something and the general level of everything. As this unity between being and non-being, possibility is a relationless, indeterminate receptacle, open to the being, the nothing, and the becoming of anything and everything, of any content whatsoever.

But there is one exception. Something cannot both be itself and the contrary of itself in the same time, manner, or place.[16] In this way, Hegel begins from the classic definition of possibility as whatever does not entail contradiction.[17]

But this exception causes a conflict. Hegel has already explained, at premise five, that if something is possible, it can and can not be, claiming that possibility itself contains contrary actualities. This is the problematic of formal modality. On the one hand, everything is possible that does not entail contradiction. On the other hand, the possible itself is the actual in its contrariety, both the actual and the opposite of this as one unity. How can possibility be both contraries unless it entails contradiction?

Commentators have offered a great deal of explanation to justify Hegel's seemingly paradoxical claim that "everything is possible" and "everything is just as much something contradictory and, because of this, impossible." The main branch of this interpretation comes from Burbidge, who suggests that Hegel means everything together is impossible, stressing the universality of everything, while at the same time "everything" is possible, stressing the mere possibility of each particular thing (SC 19). Stekeler-Weithofer makes a similar observation when he calls this passage a "title sentence," objects that Hegel's choice of words makes the passage sound more paradoxical than it is, and then, like Burbidge, distinguishes between the impossibility of everything and the possibility of each individual thing (MN 222-3). McTaggart also says that Hegel's language is misleading. On McTaggart's account, Hegel means that possibility is only impossible if it contains no reference to actuality whatsoever.[18] While I think this branch of commentary is plausible, I also think that Hegel actively intends the passage to be paradoxical. If we explain why the passage is not paradoxical, we lose much of what is important about Hegel's modal argument. In this respect, Lampert really interprets Hegel well when he emphasizes how polemic the passage is, arguing that "the function of a possibility is to express the totality, but that no one possibility can express everything the totality expresses without generating contradictions. Each possibility thus fails to express all that it *itself* expresses" (FM 75).

To investigate this debate, it will help to understand what Hegel means by limitless multiplicity. He claims that limitless multiplicity results from the contrariety of possibility, that what is possible can be and also can not be. But although it is limitless, this multiplicity retains a certain determination because of contradiction. Contradiction divides possibility into a multiplicity. This multiplicity is both limitless and the limited. It is limitless in the sense that formal possibility is like an empty container that holds all sorts of things together without changing the content of what it contains. It is also limited in the sense that the diversity of possibility is a determinate diversity set against opposition. How can the limitless and the limited come from the same source? This is another version of the question, how can possibility be both contraries unless it entails contradiction?

It will also help to understand in what sense non-contradiction restricts possibility. We should be careful not to conflate the negative moment of possibility with impossibility. The difference between the can be and the can not be of possibility is not the difference between possibility and impossibility (the cannot be). Aristotle anticipates this conflation in *De Interpretatione* (CWA 21a34–22a12) when he explains that the opposite of "it is possible for something to be" is not "it is possible for something not to be," but rather "it is not possible (impossible) for something to be." If somebody has the capacity to walk, Aristotle points out, this person can choose to walk or not to walk. It would be a mistake to say that when this person chooses not to walk, this person cannot walk. There is a difference between the contrariety of the possible and the contrary of the possible. What is possible is of contrariety in the sense that it both can and can not be. But the contrary of the possible is the impossible, the cannot be. Possibility itself contains both the yes and the no of actuality. The yes and the no are one unity, and it is this unity of being and non-being that actuality actualizes. Impossibility, on the other hand, is the opposite of possibility, a seizure or erasure of being and non-being.

When Hegel concludes that "everything is just as much something contradictory and, because of this, impossible," we can assume that he means everything is possible but that "the" everything, if taken together as one actuality, as Burbidge and others explain, is impossible since this would mean the actualization of contraries, which would be contradictory. We can assume Hegel does not mean that each individual thing is just as certainly impossible as it is possible. However, I think it is still important to keep this other textual interpretation in mind. While each actual-existent is possible and non-contradictory, each is possible because the contrary of the existent is equally possible. In this sense, possibility posits diversity as the expression that it ought to be "the" everything.[19] Since actualization cannot maintain "the" everything, what becomes actual has against it a diversity of others. Opposition thus occurs as the inner-structure of the diversity because the positive and the negative moments of possibility are both actual-existents but cannot maintain themselves together as one unity. This opposition, in turn, becomes contradiction because the diversity is just an expression of possibility as the reflection of actuality-into-itself. In other words, the opposition-diversity of something set against another something is an external expression that possibility ought to be the totality of form, and that as the totality of form, possibility is the indivisible unity of A and -A.

Lampert offers an excellent reading of possibility as the "totality of form" when he claims that things are only possible if they can express the totality, and that things are not possible if they cannot do this (FM 75). If Hegel means that there are only certain types of content that can express the totality, this would complicate his initial point about boundless multiplicity, that everything and anything is possible as long as there is no contradiction. But if Hegel only means that the form must be such that what is possible expresses the totality, one can still recognize boundless multiplicity in this.

The result, then, is not simply that the actual A contains the actual -A in its own concept (in the way that the A and the -A are contained in possibility), but rather that if the actual A is possible, it is the complete totality of "the" everything, and it is really this totality that makes it possible as one instance of itself. Obviously, something is what it is and is not what it is not. The horse in the barn is a horse and not a cow or a tree. The horse is not both a horse and not a horse. This would make no sense. However, what is equally obvious is that this particular horse is the actuality of possibility, and that as possibility, the horse carries the form of totality and not only a moment of the totality. This is why actuality in the larger sense of Hegel's meaning ("actuality" is the name of the section as well as the chapter) is the transposition of existence and essence together as one unity.

7. Actualization cannot maintain this contradiction of the possible as the impossible.

> Als diese Beziehung aber, daß in dem einen Möglichen, auch sein anderes enthalten is, ist sie der Widerspruch, der sich aufhebt. Da sie nun ihrer Bestimmung nach das Reflectirte, und wie sich gezeigt hat, das sich aufhebende Reflectirte ist, so ist sie somit auch das Unmittelbare, und damit wird sie Wirklichkeit. (WL 204)

> This relation, in which the possible also contains its other, is a contradiction that sublates itself. Now, since its determination is reflection, as we have seen, a reflection that sublates itself, possibility is therefore also the immediate and, as such, becomes actuality.

Since possibility contains both the identity of the existent-actual and the contrary of the existent-actual, and contains both equally as one unity, everything is possible that does not contradict itself; however, the totality of everything is just as much something impossible and self-contradictory. This is the case because possibility harbors self-contradiction within the function that it serves as actuality's identity-with-self. This is why Hegel claims provocatively that "therefore possibility is a self contradiction all by itself, in other words, it is the impossible" (WL 204).

To avoid this contradiction, common sense thinking attempts to separate possibility from actuality. Common sense thinking dismisses the contradiction and its underlying productivity by establishing axiomatic definitions for actuality and possibility.[20] If actuality and possibility are each different concepts, they do not then transition into each other completely but maintain only enough minimal contact to produce superficial conversions. Common sense thinking thereby covers over the *aporia* that would otherwise lead, as the complete transference of actuality and possibility into each other, to the contradiction of actualizing across contraries.

But Hegel assumes no such separation. An actual is a possible, and this produces not just minimal conversions but rather the constitutive contradiction of an actual that is at the same time the totality of possibility. Evidence for this interpretation comes by way of Hegel's claim that what is essential of actual existence is not something separate from actual existence, but rather exists just as it does, as a moment of its essence. That which calls actuality into being, and stands against it as its true essence, is just the process of actuality becoming itself. If actuality is possibility in this way, then actuality is its own other in the radical sense that the negative is its own positive. This makes possibility not just the relating ground between A and -A. This makes possibility the A and the -A as actuality's own self movement.

Hegel concludes at this point in the argument that possibility is self-contradiction. This conclusion causes Hegel to revise what actuality and possibility mean. Actuality becomes the truth of possibility and the extensive openness of possibility becomes an existent-actual. Because of the importance of the bind between actuality and existence at premise one, Hegel calls this revision "reflected actuality."

8. Because of this contradiction, actuality becomes reflected actuality. Reflected actuality is an actuality of possibility itself.

> Diese Wirklichkeit ist nicht die erste, sondern die reflectirte, gesetzt als Einheit ihrer selbst und der Möglichkeit. Das Wirkliche als solches ist möglich; es ist in unmittelbarer positiver Identität mit der Möglichkeit. (WL 205)

> This actuality is not the initial actuality, but is rather reflected actuality, posited as the unity of itself and possibility. The actual as such is possible; it has an immediate, positive identity with possibility.

Before he turns to contingency, which accounts for a disjunctive division between actualized possibility and unactualized possibility, Hegel first entertains what he calls reflected actuality: to let thought attempt to think of actuality as possibility itself. Reflected actuality reveals that his initial theory of simple actuality (premise one) is insufficient because it does not explain the negativity of the possible. The sides of possibility cannot simply become one actuality, nor can the possible become divided in the actual, without becoming merely a fraction of the complete totality. This is the state of affairs that leads Hegel to deduce reflected actuality. Reflected actuality is an actuality of the possible itself. It is an actuality that is not actuality at all, but is only the empty form of positedness in an attempt to express the possible *qua* the possible.

Most commentaries skip over the reflected actuality steps of the argument, perhaps justifiably, since Hegel has not properly emphasized the role that these steps play in the chapter, and since from the outset reflected actuality looks like an inadequate and unsustainable position. Reflected actuality is nevertheless an important transition-point for Hegel. By attempting to establish an actuality that is possibility itself and by explaining why this does not work, Hegel is able to re-expose the contradiction at premise seven from another disposition. He is also able to discuss possibility *qua* possibility and the possible as the unactualizable.[21]

9. The problem with reflected actuality is that since it is the possible itself, it is an actual that cannot become actual.

> Aber [reflectirte Wirklichkeit] hat sich bestimmt als nur Möglichkeit; somit ist auch das Wirkliche bestimmt als nur ein Mögliches. Und unmittelbar, darum weil die Möglichkeit in der Wirklichkeit unmittelbar enthalten ist, ist sie darin als aufgehobene, als nur Möglichkeit. Umgekehrt die Wirklichkeit, die in Einheit ist mit der Möglichkeit, ist nur die aufgehobene Unmittelbarkeit; - oder darum weil die formelle Wirklichkeit nur unmittelbare erste ist, ist sie nur Moment, nur aufgehobene Wirklichkeit, oder nur Möglichkeit. (WL 205)

> Reflected actuality has determined itself to be only possibility; therefore, the actual is only a possible. Because possibility is immediately contained in actuality, it is contained in it as sublated, as mere possibility. Conversely, actuality, which is in unity with possibility, is only sublated immediacy; or, in other words, because formal actuality is only the initial, immediate actuality, it is only a moment, only the sublated actuality, in other words, only possibility.

Try to think A and -A as one unity. This is impossible. Even the pure conjecture of the imagination cannot visualize A and -A together without transposition. Certainly, I can transpose A and -A. But transposition requires a distinction of time, manner, or place, a disjunction of the actual in the possible. To attempt to render in actuality the immediate status of the possible *qua* the possible is to present only the indeterminateness, incompleteness, and vagueness of the possible that has no relation to the actual. This is why reflected actuality fails. While the possible A contains the possible -A, immediate actuality cannot present this with any distinctness. Reflected actuality is the actuality of the unactualizable and the unpresentable.[22] What appears in actuality is the total absence, the nothing, of the possible that has no further relation to the actual.

We can anticipate (many steps ahead of Hegel's deduction) that the actualization of the possible *qua* the possible will require mediation. While thought cannot think A and -A together as one unity, there are ways through mediation to integrate A and -A together into one unity. We can also anticipate why Hegel must turn from form to content: because actualization cannot actualize the possible unless it finds a way to complete itself in the negative of itself, through the mediation of itself as content, by integrating its other as what it is.

Remark: Kant on the Agreement of the Possible and Experience

In *The Critique of Pure Reason* under the heading the "Postulates of Empirical Thinking in General," Kant thematizes the modal categories by claiming that what is possible is that which stands in agreement with the formal conditions of experience.[23] Possibility does not extend beyond experience, but rather exists as the coincidence of the actual and agrees with it in every case. To this end, Kant defines all three modal categories (leaving contingency out of his analysis) in terms of experience:

1. Whatever agrees with the formal conditions of experience (in accordance with intuition and concepts) is possible.
2. That which is connected with the material conditions of experience (of sensation) is actual.
3. That whose connection with the actual is determined in accordance with general conditions of experience is (exists) necessarily (CPR 321).[24]

Although it might seem plausible to assume that possibility does not need to agree with experience, in the sense that many things do seem possible that are not actual, Kant claims that this notion of mere possibility which would seem to operate beyond any experience whatsoever leads only to the dispositions of dogmatic metaphysics. To assert that possibility does not need to agree with actuality brings us beyond the transcendental conditions for the possibility of experience, whether this means the realist who attempts to investigate what things are really like beyond the subject's experience, or the material idealist (such as Kant attributes to Descartes and Berkeley) who rejects *a priori* knowledge claims as either indemonstrable or impossible.[25] Attempts to recognize possibilities that are not grounded in actuality present us only with what Kant calls "figments of the brain," which, while they are not formally self-contradictory, cannot establish the necessary *a priori* conditions of experience in terms of possibility. Mere possibilities appear only through the conflation of the *a priori* with the *a posteriori*. From within experience, many things seem to be possible that exist beyond any agreement with actuality.[26] But Kant goes on to claim that the modes of modality require the subject's experience,[27] not in order to expose empirically how things could have been otherwise, but to evaluate the necessary modes of modality, which do not exist in the constitution of things, but only in the subjective synthesis (CPR 332).

10. Hegel turns instead to contingent actuality, where the actual as what is immediately given posits its other as what could have been.

> Alles Mögliche hat daher überhaupt ein Seyn oder eine Existenz. (WL 205)

> Diese Einheit der Möglichkeit und Wirklichkeit ist die Zufälligkeit.—Das Zufällige ist ein Wirkliches, das zugleich nur als möglich bestimmt, dessen Anderes oder Gegentheil eben so sehr ist. Diese Wirklichkeit ist daher blosses Seyn oder Existenz, aber in seiner Wahrheit gesetzt, den Werth eines Gestztseyns oder der Möglichkeit zu haben. Umgekehrt ist die Möglichkeit als die Reflexion-in-sich oder das Ansichseyn gesetzt ale Gesetztseyn; was möglich ist, ist ein Wirkliches es in diesem Sinne der Wirklichkeit, es hat nur so viel Werth als die zufällige Wirklichkeit; es ist selbst ein Zufälliges. (WL 205)

> Everything possible has in general being or existence.

> This unity of possibility and actuality is contingency. The contingent is an actuality that is, at the same time, determined only as possibility, whose other or opposite equally is. This actuality is merely being or existence, but is posited in its truth as having the value of positedness or as having the value of possibility. On the other hand, possibility is self-reflectedness, in other words, the in-itself posited as positedness. What is possible is an actual in the sense of actuality. It has the same value as contingent actuality. It is itself something contingent.

One of the more controversial statements of the chapter comes from Hegel's claim "everything possible has . . . being or existence" (*Alles Mögliche hat . . . Existenz.*) If by this Hegel literally means that all possibilities entail themselves in actuality, the world that this describes would be quite absurd. We would be unable to distinguish what exists from what is merely possible. Mere possibility would mix together with existence and cause a world of surrealist specters and dreams. As the modal logicians point out, while actuality entails possibility, possibility does not necessarily entail actuality. In *the Encyclopaedia Logic*, Hegel also expresses the same reservation about mere possibility, emphasizing the absurdity of the immediate entailment from possibility to actuality, when he writes,

> So kann auch das Absurdeste und Widersinnigste als möglich betrachtet werden. Es ist möglich, daß heute abend der Mond auf die Erde fällt, denn der Mond ist ein von der Erde getrennter Körper und kann deshalb so gut herunterfallen wie ein Stein, der in die Luft geschleudert worden; - es ist möglich, daß der türkische Kaiser Papst wird, denn er ist ein Mensch, kann als solcher sich zum Christentum bekehren, katholischer Priester werden usw . . . möglich sei dasjenige, wofür sich ein Grund angeben lasse.[28]

> [E]ven the most absurd and nonsensical suppositions can be considered possible. It is possible that the moon will fall on the earth this evening, for the moon is a body separate from the earth and therefore can fall downward just as easily as a stone that has been flung into the air; it is possible that the Sultan may become Pope, for he is a human being, and as such he can become a convert to christianity, and then a priest, and so on. . . . Anything for which a ground (or reason) can be specified is possible. (EL 216)[29]

The statement "everything possible has a being or existence" is less problematic, however, if we interpret Hegel to mean that what actually exists posits the alternative of itself, and that when we take into account this state of affairs, everything possible has existence. In this sense, immediate actuality posits the equal existence of alternative possibilities, and with this the complete form of the possible itself, through the expression of contingency.

It is from these terms that Hegel turns to formal contingency as the resolution of the problem of reflected actuality. Whereas reflected actuality is an actuality of the merely possible, contingency is an actuality that is truly actual, but whose opposite equally exists. In contingency, the negativity of the possible—that what is possible can and can not be—becomes contingent actuality and alternative possibility. Although this actual is, its other could have been. Contingent actuality more effectively renders the status of possibility itself because the can not of possibility appears through whatever happens immediately to exist in actuality, and in this way, possibility really does exist. If A is a contingent actuality, then A posits -A as what equally exists. Contingency thus presents, in a certain respect, the actual A as containing the existence of -A within its own concept. The reason why this is no longer a contradiction is because A and -A contain each other in a relationship of indifference. If A is actual, this posits -A alongside it as what could have been.

Let us visualize what Hegel has in mind. The actual world appears in its immediate givenness. This is the simple actuality of what is already there. The horse, for example, is there in the barn. The other swimmer is there in the lane. If I look at a map of the earth, I see the mountains and the lakes as already there, as what is simply given to their region. Yet, since this that just appears before me is itself possibility, I recognize in what is already there something more than what is already there.

This leads me to two further points. I recognize alongside the actual-existent that the possibility of its opposite has of itself an existence, and that this existence could have been what is actual. But I also recognize not only that possibility exists, but also that the immediately existing actual only happens to be. While what immediately appears has the authority of truth, it appears at the same time as finite, since it appears with its other alongside it. Hegel does seem to suggest at this point that contingency is the only possibility for actuality, in the sense that the opposite of what is always appears alongside the simple givenness of what is. But since he does not yet determine that this other who is posited alongside must overtake this simple actuality, the connotations of finitude extend us further than what Hegel suggests at this point in the text.

Hegel is not only saying that in the contingent actual, the other appears too, and that these sides together express the possible itself. He is also saying that actuality depends upon the equal existence of the possible as what could have been. The swimmer passes me in the lane; yet, the quality of this simple event is formed in the contingency and instability that what happens could not have happened. I see in the body movement of the other swimmer not only the appearance of what could have been but the literal texture of the possible existing in the actuality of what is. Hegel will soon make this point again in the transition from formal contingency to formal necessity, when he argues that although A contains -A as an existent, it depends upon the non-actuality of -A, because otherwise A cannot be actual. But before Hegel exposes this transition, he argues that what is contingent is both grounded and groundless.

11. If the other of the actual equally exists, there is no reason why this actual is and why its other is not. Therefore, contingency has no ground.

> Das Zufällige bietet daher die zwey Seiten dar; erstens insofern es die Möglichkeit unmittelbar an ihm hat, oder, was dasselbe ist, insofern sie in ihm aufgehoben ist, ist es nicht Gesetztseyn noch vermittelt, sondern unmittelbare Wirklichkeit; es hat keinen Grund.—Weil auch dem Möglichen diese unmittelbare Wirklichkeit zukommt, so ist es so sehr als das Wirkliche, bestimmt als zufällig, und ebenfalls ein Grundloses. (WL 205–6)

> Contingency therefore presents two sides. First, because it has possibility immediately in it—or, what is the same thing, because possibility is sublated in it—contingency is neither something posited nor mediated, but is, instead, immediate actuality; it has no ground. Because this immediate actuality also belongs to possibility, the latter no less than the actual is determined as contingent and likewise as *groundless*.

There is no distance in reflected actuality between actuality and possibility. Contingent actuality corrects this problem by uniting possibility as an actuality whose opposite equally exists. However, contingency causes a further problem. An actual whose opposite equally exists suffers from indifference. Contingency combines immediate actuality with the equal existence of other possibilities; if the contingent were to make the difference between these two modes explicit, then either we would have to retain the initial contradiction at premise seven, where actuality is both the can and the can not as one actuality, or we would have to retain the problematic position of reflected actuality, where the actual is merely a possible and otherwise lacks itself. It is only through indifference that contingency sustains the actuality of possibility.

The claim from contingency shows that the other of the actual is as present to the being of what is as its own actuality is present and immediate. The actual itself only is because its opposite is not. Equally, the actual only is because its other could have been. Contingent actuality therefore appears as groundless, and in this way reveals the existence of possibility, no longer as an omission or remainder that escapes the actual, but now as the groundlessness of what is immediately actual. Hegel's point is not merely that the actual is arbitrary, in the sense that the opposite of what is could just as well have been. Hegel's point is that actuality expresses the existence of possibility only through its contingency.

At this initial stage, Hegel's use of the word contingency (*das Zufälligkeit*) is quite similar to a conventional definition of contingency, that if something is contingent, then it could have been otherwise. This is the opposite of a conventional definition of necessity, that if something is necessary, then it could not have been otherwise. But Hegel will show that the same process that makes contingency groundless also makes the actual grounded in its other.[30]

12. But if the other of the actual equally exists, then actually depends upon what could have been.

> Das Zufällige ist aber zweytens das Wirkliche als ein nur Mögliches oder als ein Gesetztseyn; so auch das Mögliche ist als formelles An-sich-seyn nur Gesetztseyn. Somit ist beydes nicht an und für sich selbst, sondern hat seine wahrhafte Reflexion-in-sich in einem Andern, oder es hat einen Grund. (WL 206)
>
> However, second, contingency is the actual as something only possible, in other words, as positedness; therefore, the possible is also as the formal in-itself only a positedness. Consequently, the two are both not in and for themselves but each has its self-reflection in an other, or each has a ground.

Hegel then says the exact opposite about contingency. Whereas the first argument from contingency presents an actual whose immediate existence rests in the groundlessness that what is could have been otherwise, the second argument from contingency claims, to the contrary, that the actual is only actual if its other is not. This second argument exposes the contingent and seemingly groundless actuality as really grounded in the prior conditions of its other.

Hegel's argument at premise twelve invokes a classic definition of contingency, current in medieval modal theories, that what is contingent is caused by a condition that is prior. Contingency is grounded in the sense that what exists only becomes actual when its conditions of possibility are met.

The other that exists alongside the existent-actual is integral to the process by which something comes to emerge into actuality in the first place. This means that contingency is not only the indifference of the actual and the opposite of the actual. It is also the sign that what is immediately given has emerged into actuality from a prior source. This is nevertheless a concept of contingency for Hegel because it shows that what is simply given is not just the surface of the actual but also the prior and subsequent existence of something else. What shows through the surface is the appearance of the process of actualization. The mountains and the lakes, while they appear as simply given to their region, reveal in their contingency that the earth must have come to form them from a position prior to their actuality. In this sense, the "contingent upon another" version of contingency exposes the possible *qua* the possible, not only by including what could have been, but by exposing the possible as what the actual has become.

Burbidge's contribution to the analysis of Hegel's modal concepts is especially pronounced when it comes to this transition between contingency and necessity in Hegel's formal argument. Generally, Burbidge claims that the securing of contingency in necessity is the main conclusion of Hegel's argument in the "Actuality" chapter. There are three versions of this conclusion in Burbidge's analysis, the first two acting as material for the third:

(1) The formal version states that whatever happens contingently to be actual is necessarily contingent in the sense that what is must be, simply because it is. Burbidge rejects this version because it leads to what he claims is the flaw of Megarian Actualism, the argument that only what is actual exists, and that what is possible but not actual does not exist (SC 28–29).

(2) The real version explains why conditional necessity begins from contingency. "Because of real possibility *A*," Burbidge writes, "*B* must become actual. But that necessity is contingent on the specific determinations of *A*" (SC 40). We will come to this formulation of contingency and necessity at premise twenty-two.

(3) Ultimately, Burbidge claims that the disposition of totality requires contingency and that this contingency is absolutely necessary. "This is the nature of necessity when we consider the total picture—what Hegel calls 'absolute necessity'—and it requires, as a defining gesture of its complex dynamic, that there be contingencies" (SC 47). We will come to this formulation at premise twenty-seven. What we witness here between premises twelve and thirteen is the formal transition from contingency to necessity, which I think Burbidge's analysis only partially explains.

13. Formal necessity is the source of these two arguments from contingency. Formal necessity is the coincidence of actuality and possibility.

> Das Zufällige hat also darum keinen Grund, weil es zufällig ist; und eben so wohl hat es einen Grund, darum weil es zufällig ist. (WL 206)

> Diese absolute Unruhe des Werdens dieser beyden Bestimmungen ist die Zufälligkeit. Aber darum weil jede unmittelbar in die entgegengesetzte umschlägt, so geht sie in dieser eben so schlechthin mit sich selbst zusammen, und diese Identität derselben einer in der andern ist die Nothwendigkeit. (WL 206)

> Das Nothwendige ist ein Wirkliches; so ist es als unmittelbares, grundloses; es hat aber eben so sehr seine Wirklichkeit durch ein anderes oder in seinem Grunde, aber ist zugleich das Gesetztseyn dieses Grundes und die Reflexion desselben in sich; die Möglichkeit des Nothwendigen ist eine aufgehobene. Das Zufällige ist also nothwendig, darum weil das Wirkliche als Mögliches bestimmt, damit seine Unmittelbarkeit aufgehoben und in Grund oder Ansichseyn, und in Begründetes abgestossen ist, als auch weil diese seine Möglichkeit, die Grundbeziehung, schlechthin aufgehoben und als Seyn gesetzt ist. Das Nothwendige ist, und diß Seyende ist selbst das Nothwendige. (WL 206–7)

> The contingent has no ground because it is contingent, but, equally, has a ground because it is contingent.

> Contingency is the absolute unrest of the becoming of these two determinations. Yet, because each immediately turns into its opposite, equally in this other it simply unites with itself, and this identity of both, of one in the other, is necessity.

> The necessity is an actual. Because of this, it is something immediate and groundless. But at the same time, it has its actuality through an other or in its ground. Yet it is equally the being posited of this ground and its reflection into itself. The possibility of the necessity has been sublated. The contingent is therefore necessary, just because the actual is determined to be possible and so its immediacy is sublated and pushed off into ground or being-in-itself and the grounded; and also because this its possibility or grounding relation has been completely sublated and posited as being. The necessary is, and this being is itself the necessary.

Common sense conceives of necessity as the opposite of contingency. What is necessary cannot be otherwise, whereas what is contingent can be otherwise. While it certainly makes reference to these conventional definitions of necessity and contingency, what is exciting about Hegel's claim is that necessity literally comes out of contingency as the result of contingency's being both groundless and grounded. Necessity is an advancement from contingency because it takes away the indifference of content and replaces it with the true coincidence of the actual as the possible.

Since there has been a lot of debate about formal necessity, I will briefly outline what I perceive as the two most plausible theories from secondary sources. My interpretation is that these two theories must be combined together, and only then can we make sense of Hegel's passages about formal necessity.

The first theory comes from Houlgate's and Burbidge's reading of formal necessity as the explicit formalization of the law of non-contradiction. While either contrary is possible, only one or the other is actually possible. Houlgate's definition of formal necessity follows along these lines. Formal necessity is "simply the impossibility of *possibility* itself being anything other than *actual* possibility, the impossibility of possibility's being *mere* possibility" (NC 41).[31] Formal necessity marks the impossibility of grasping the totality of possibility in one actuality. It marks the limit of the possible in the actual.

However, we find in this formalization of necessity and the law of non-contradiction a certain redundancy. By premise seven, Hegel has already determined that in the most immediate sense actualization *cannot* actualize the extent of possibility without falling into contradiction. If formal necessity were only that the possible *qua* the possible cannot occur, and that possibility itself is always only the possibility of actuality, Hegel would have no reason to turn to content-modality. The solution to the problem of how to actualize possibility would simply be that thought cannot think the totality of possibility with any determinateness, but *must* always remove the contrariety of the possible so that the possible can become actual. This is the solution from disjunction. But I think this is only one of the consequences of formal necessity.

Another consequence is that there is a necessary form that actualization must accept in order to actualize across contrary possibilities. If actualization cannot actualize possibility itself in the immediate sense, as we have already explained, then actualization will need to find another path into possibility. Since the insight from contingency shows us that everything possible exists, we can conclude from this that unactualized possibilities can still become actual. Formal necessity, then, replaces the indifference of contingency (that the actual A contains the mere existence of -A alongside it) with a mode of actualization that can integrate the differences between A and -A.

This second theory follows along the same lines as the reading that Lampert offers when he defines formal necessity as "the totality out of which anything actual must be formed" (FM 75). Lampert explains this transition from contingency to formal necessity as a force that propels the actualization of the contrary sides of possibility (FM 76). Formal necessity marks not only the impossibility of actualizing, strictly speaking, the negativity of the possible, which always in formal terms turns the contrary into the contradictory. Formal necessity is also the pressure, the telos, to complete the possible in the actual. Just as the form of actualization requires the removal in actuality of the other possibility, likewise, it requires that this other side become actual through transposition and mediation. Necessity simultaneously removes the opposite of the actual while gathering what it removes and taking this as the actual as well. Necessity is thus both what takes the possibility away, in the sense that what is necessary is the possible with the contrary possible discharged, but also the force that spurs actualization to actualize the other side, and thereby the possible itself by way of mediation.

Hegel concludes that since the formal structure of self-identity does not have the resources to mediate between the negative and positive sides of possibility, we require a different model of actualization. Actuality becomes the actuality of internalizable content because only in terms of content can actualization find its possibilities in other actuals. We turn thus from the formal laws of identity and non-contradiction to the contextual world of actuality and its complex web of possibility-relations, transfers, and conditionals.

NOTES

1. Unless otherwise noted, this translation of Hegel's "Actuality" chapter is my own. I have consulted and benefited from Di Giovanni's 2010 translation in *The Science of Logic*, Burbidge's 2007 translation in *Hegel's Systematic Contingency*, and Miller's 1969 translation in *Hegel's Science of Logic*.

2. For discussions of Hegel's distinction between existence as the thing-emerging-process and the fact of existence, see Marcuse (HO 93), Di Giovanni (CC 199, endnote 18), and Houlgate (NC 38).

3. Burbidge explains this when he writes "that the actual incorporates the possible specifies its difference from the apparently synonymous terms: 'being' and 'existence'"(SC 17). Also see John W. Burbidge, *The Logic of Hegel's Logic* (Peterborough: Broadview Press, 2006), 75.

4. Marcuse says that "what is actual is possible" is the first premise of Hegel's argument. But I recognize "what is actual is existence" first because this establishes Hegel's starting point in actuality, not possibility. Marcuse is still right to point out that the argument only begins to make sense as an argument once Hegel introduces the relationship between actuality and possibility (HO 93).

5. If the claim "actuality contains possibility" implies that actuality is logically prior than possibility, then we can assume that there is an affinity between this passage from Hegel and Aristotle's various accounts of the logical and ontological priority of actuality. For Aristotle's complex and in some ways ambiguous arguments for the primacy of actuality, see division eight of *Metaphysics* Theta (especially CWA 1049b4–1050a16). Also see my discussions of this in Nahum Brown, "The Modality of Sovereignty: Agamben and the Aporia of Primacy in Aristotle's *Metaphysics* Theta," *Mosaic* 46, (2013): 169–82 and Nahum Brown, "Aristotle and Heidegger: Potentiality in Excess of Actuality." *Idealistic Studies* 46.2 (2017): 199–214.

6. Hegel does not mention experience (*Erfahrung*) in these passages of the "Actuality" chapter because this is not how the *Logic* defines its subject matter. Nevertheless, it would be worthwhile to anticipate the logical analysis he sets up in the "Actuality" passages of the *Logic* with the "Science of Experience" project of the *Phenomenology*. Two excellent books that analyze the *Phenomenology* as a science of experience are H. S. Harris, *Hegel: Phenomenology*

and System (Indianapolis: Hackett Publishing, 1995) and John Russon, *Reading Hegel's Phenomenology* (Bloomington: Indiana University Press, 2004).

7. Melvin Fitting and Richard L. Mendelsohn, *First-Order Modal Logic* (Dordrecht: Kluwer Academic Publishers, 1998), 5. For an explanation of the "axiom of possibility" and the related "axiom of necessity," see G. E. Hughes, and M. J. Cresswell, *A New Introduction to Modal Logic* (London: Routledge, 1996), 28. "If p, then possibly p" holds only in systems with reflexivity built into them, such as S4 and S5, but does not hold in systems such as K4, where the only frame condition is transitivity. Systems without reflexivity, such as K (no frame conditions), D (only the serial condition), and K4 (only the transitivity condition), cannot establish "if p, then possibly p" because there may not be access to the actual world. In *First Order Modal Logic*, Fitting and Mendelsohn discuss this in terms of the related necessity axiom, "if p is necessary, then p is actual," *First-Order Modal Logic*, 9–10.

8. Longuenesse discusses premise five (possibility is the comparing relation between A and not A) but in a different order. She attempts to place premise five after premises six and seven, as if Hegel's definition of possibility as comparing relation were the result of the contradiction of actualizing A and not A as one unity (CM 125). On my reading, premise two (what is actual is possible) and premise five (possibility is both A and not A) infer the contradiction at premise seven (actualization cannot actualize the contrariety of possibility without contradiction), and this impossibility of actualizing the totality results in formal contingency at premise ten and formal necessity at premise thirteen.

9. Cf. Brouwer's theory of Intuitionism, which criticizes the assumption that there is an equivalency between the principle of identity and the principle of double negation, the law of the excluded middle, and the principle of non-contradiction. For his account of the debate, see L. E. J. Brouwer, "Intuitionism and Formalism," *Bulletin of the American Mathematical Society* 37 (1999): 55–64.

10. "On What We Can Not Do" in Giorgio Agamben, *Nudities*, translated by David Kishik and Stefan Pedatella (Stanford: Stanford University Press, 2011), 44–45 contains one of Agamben's most insightful contributions to theories of potentiality, but there are many other contributions as well. The "Potentiality and Law" chapter of *Homo Sacer* offers an ontological-modal basis for the political sovereign paradox, in the form of an analysis of Aristotle's *Metaphysics* over the constitutive ambiguity between whether actuality or potentiality is more primary. Agamben also emphasizes the importance of the can not in various related concepts throughout his corpus. It appears as the concept of the "whatever" in Giorgio Agamben, *The Coming Community*, translated by Michael Hardt (Minneapolis: University of Minnesota Press, 1993), 9. It appears as the relation between example and exception in Giorgio Agamben, *Homo Sacer: Sovereign Power and Bare Life*, translated by Daniel Heller-Roazen, (Stanford: Stanford University Press, 1998), 21–23. It appears as inoperativeness and openness, and in the "As If" and "Exigency" sections of Giorgio Agamben, *The Time That Remains: A Commentary on the Letter to the Romans*, translated by Patricia Dailey (Stanford: Stanford University Press, 2005). It also appears prominently in his collection of essays: Giorgio Agamben, *Potentialities: Collected Essays in Philosophy*, translated by Daniel Heller-Roazen (Stanford: Stanford University Press, 1999). Leland De la Durantaye, a major commentator of Agamben, goes so far as to call Agamben the philosopher of potentiality. Leland De la Durantaye, *Giorgio Agamben: A Critical Introduction* (Stanford: Stanford University Press, 2009), 4.

11. Agamben seems to conflate the terminology between the impotential and the potential not to be, that is, between the cannot and the can not. For example, Agamben writes, "[Today's man] has become blind not to his capacities but to his incapacities, not to what he can do but to what he cannot, or can not, do." By incapacities, Agamben means the capacity not to be. However, it is not clear whether his equation of the two terms is for some reason intentional. But since he does not conflate the conceptual distinction, but in fact maintains this distinction consistently throughout the essay, whether he writes "cannot" or "can not" is only a minor distraction to his argument. For an explanation of the conceptual difference between the cannot and the can not, see Aristotle's *De Interpretatione* (CWA 21a34–22a12), and my discussion of this at premise six.

12. Agamben likes to illustrate this strategy through Melville's character Bartleby. Bartleby effectively suspends the potential from its end in actualization by employing the modal unde-

cidability of the "I would prefer not to." See Agamben, "Bartleby, or on Contingency" in *Potentialities*. Also see the "Potentiality and Law" chapter of Agamben, *Homo Sacer*.

13. Agamben, *Potentialities*, 181.

14. Agamben, *Nudities*, 44.

15. Agamben, *Nudities*, 45.

16. For Aristotle's discussion of the principle of non-contradiction, see Book IV of the *Metaphysics*. At the end of division three, Aristotle says that the principle of non-contradiction is the most evident of all principles. He defines non-contradiction in terms of impossibility when he writes, "It is impossible that contrary attributes should belong at the same time to the same subject" (CWA 1005b26–27). Kant also discusses the principle of non-contradiction in terms of impossibility (CPR 279–81). He presents the principle of non-contradiction as the analytic judgment version of the "System of the Principle of Pure Understanding." His account of modality then follows from this as the fourth principle of the synthetic judgments. In *A Study of Hegel's Logic*, Mure situates Hegel's point about possibility and non-contradiction in terms of Aristotle, Kant, and also Leibniz. See, 134–36.

17. Houlgate points out that because it is the identity of actuality into itself, possibility equally exposes the non-contradictoriness of -A= -A. "Surely, the possibility of not-A rests just as much on non-contradictoriness (namely, that of not-A) as does the possibility of A" (NC 39–40).

18. McTaggart, *A Commentary on Hegel's Logic*, 164.

19. Hegel calls possibility the ought-to-be of the totality of form at WL 203-4.

20. By assuming axiomatic definitions of modal terms, modal logicians thereby avoid the contradiction at premise seven. Conventionally, if a proposition is possible, this means that it can be true or false, but that it is not necessarily false. If a proposition is necessary, it cannot possibly be false. And if a proposition is merely contingent, it can be either true or false, but neither necessarily. However, how the modal logicians come to these axioms can be viewed as problematic. Stekeler-Weithofer claims that by assuming definitions from the outset, "formal modal logic misses our real practice of talking about real possibilities and objective reality" (MN 219). Stekeler-Weithofer says that for Hegel reality depends on the real existence of possibility. Modal logicians cannot approach modal reality in a meaningfully existential way because of the axiom assumptions that come along with possible worlds semantics.

21. Some commentators do discuss reflected actuality from slightly different terms. Burbidge includes a short analysis of reflected actuality but since he sees reflected actuality, as already disjunctive (SC 22), his analysis does not uphold, even in its temporary problematic status, an actuality of the possible itself. Houlgate discusses the closely related term "suspension," but does not discuss reflected actuality directly in his essay (NC 40). Di Giovanni begins from the Kantian stance that an object is either actual or possible. But in doing this, he changes the parameters of the debate from the disjunction in actuality of possibility (that is, if possibly I swim, then either actually I swim or actually I do not swim), to the division of an object into one of these two modes—actuality or possibility (CC 182–83, 195). Di Giovanni's reliance on the thesis that an object is either actual or possible does not allow him to fully anticipate one of the most important transitions in the chapter, that it would be a contradiction to actualize both contraries of possibility.

22. The notion that possibility is impossible appears in Derrida's 1968 essay "Différance." Spelled with an "a" rather than an "e," *différance* marks off in the negative the unpresentable middle voice between the positive existence of one difference from another. Jacques Derrida, *Speech and Phenomena*, translated by David B. Allison (Evanston: Northwestern University Press, 1973), 129–60. Hegel's reflected actuality expresses a similar motif. Possibility appears only in the negative as the middle voice between one actuality and another. Every rendering of the actual in the possible is a decision that could never completely anticipate what possibility is. In this sense, possibility itself is impossibility. And yet what appears in the withdrawing of the possibility from every decision is the undecidability of the decision, the presence in the negative of the possible that will not come. For an excellent study of Derrida's essay, see Hugh J. Silverman "Self-Decentering: Derrida Incorporated" in *Inscriptions* (Evanston: Northwestern University Press, 1997), 294–315.

23. For an extensive, book-length commentary on Kant's modal theory, see Nicholas F. Stang, *Kant's Modal Metaphysics* (Oxford: Oxford University Press, 2016).

24. Cf. Kant's modal definitions in his description of the twelve categories. In the first table of categories, Kant lists possibility under the "problematic," actuality under the "assertoric," and necessity under the "apodictic." Here he says that while the categories subsumed under quantity (the universal, particular, and singular), quality (the affirmative, negative, and infinite), and relation (the categorical, hypothetical, and disjunctive) each contribute to the concept of the judgment, the categories subsumed under modality are different in that they do not contribute anything to the content of the object, but rather effect only the copula of the judgment. Although Kant defines possibility as arbitrary, contrasting this with the actual, which he defines as truth, he also recognizes that some merely problematic judgments, which are themselves false, nevertheless contribute to the "conditions of the cognition of truth" (CPR 209) since by giving an account of false paths, possibility can help to orient us to the truth (CPR 206–10).

25. See Kant's "Refutation of Idealism," included in the B edition as part of "The Postulates of Empirical Thinking in General" (CPR 326).

26. Three examples that Kant offers of this kind of mere possibility include something that is between matter and a thinking being, such as that which is present in space but which does not fill it, someone who can see the future and not only deduce it, and someone who can read others' minds (CPR 324).

27. Kant revolutionized conceptions of modality, as Longuenesse puts it, because he "no longer defined it from the point of view of God, but from the point of view of the cognizing subject" (CM 161).

28. Hegel, *Enzyklopädie der philosophischen Wissenschaften im Grundrisse*, 283.

29. For further analysis of this famous passage from *the Encyclopaedia Logic*, see Longuenesse (CM 126) and Ng, "Hegel's Logic of Actuality," 8.

30. For a related discussion of two meanings of contingency in Hegel, see Raoni Padui, "The Necessity of Contingency and the Powerlessness of Nature: Hegel's Two Senses of Contingency," *Idealistic Studies* 40.3 (2010): 243–55. Padui claims that in the *Logic* contingency primarily has the connotation of "dependence and conditionality," in contrast to the *Philosophy of Nature*, where contingency primarily has the connotation of "irrationality and chance." However, since contingency as "groundlessness" is one of the two sides of formal contingency, as Hegel outlines the distinction in the "Actuality" chapter, Hegel's use of contingency as "irrationality" and "chance" is also represented in the *Logic*. In other words, the sense of contingency that Padui attributes to the *Philosophy of Nature* is already part of Hegel's dialectical account of how contingency is both grounded and groundless.

31. Also see Burbidge (SC 28).

Chapter Two

Real Modality

14. Real actuality results from the necessary form that actualization takes to actualize possibility.

> In dieser formellen Nothwendigkeit ist daher die Einheit zunächst einfach und gegen ihre Unterschiede gleichgültig. Als unmittelbare Einheit der Formbestimmungen, ist diese Nothwendigkeit Wirklichkeit; aber eine solche, die, weil ihre Einheit nunmehr bestimmt ist als gleichgültig gegen den Unterschied der Formbestimmungen, nemlich ihrer selbst und der Möglichkeit, einen Inhalt hat. Dieser als gleichgültige Identität enthält auch die Form als gleichgültige, d. h. als bloß verschiedene Bestimmungen, und ist mannichfaltiger Inhalt überhaupt. Diese Wirklichkeit ist reale Wirklichkeit. (WL 207–8)

> The unity in this formal necessity is at first simple and indifferent to its differences. As an immediate unity of formal determinations, this necessity is actuality, but of a kind which—because its unity now is determined to be indifferent with reference to the difference of the formal determinations (that is, of itself and possibility)—has a content. As an indifferent identity, this content also contains the form as indifferent—as merely diverse determinations—and is in general a multifarious content. This actuality is real actuality.

Hegel describes the transition from form to content as the transition from one type of indifference to another. I think by the term "indifference" (*Gleichgültigkeit*) in this context Hegel means a type of relation that allows for otherwise conflicting elements to stand beside each other, without falling into direct conflict or even contradiction. Indifference allows for subsistence. To be indifferent is to put opposition and mutual exclusivity aside so that conflicts and disharmonies are momentarily suspended. The necessary form that actualization takes to actualize possibility is one that is indifferent to "the difference of form determinations." This is a significantly new type of indifference from the initial indifference of formal actuality. Formal actuality is indifferent to content differences. It is just the simple coincidence of something as identical-with-itself. But real actuality is indifferent to the form determinations, and this in turn opens actuality to the differences of content. Since Hegel has already used the term indifference to account for how contingent actuality relates to the existence of its other, it will be worthwhile to clarify these various usages.

When he writes, "as long as we hold to this simple form, the content remains something identical with itself and therefore something possible" (see premise five), Hegel means form as indifferent to content. If we do not take into account the differences of content, we remain at the merely formal level, where, if something is possible, then it is coherent-with-itself (truth-affirming). Basic deductive symbolic logic also functions in this way. We stipulate that the rules of the truth-functional connectives are indifferent to content differences, in the sense that only the form of the connective matters, but the actual content does not matter. If we take the form of the disjunctive connective "p or q," we can set up simple deductions such as these:

p or q
not p

Therefore q

But from this disjunctive form, we can infer conclusions that are inappropriate to the content. For example:

Either apples are oranges or bananas
Apples are not oranges
Therefore apples are bananas

In this sense, everything is possible that does not contradict itself. As long as we remain at the purely formal level, the content itself does not matter. Although the form of the disjunctive syllogism is valid, its conclusion is obviously inconsistent with the actual world.

Hegel's insight at premise fourteen, then, is to recognize an actuality of content that is indifferent to form. What does it mean to be indifferent to form determinations? This is to set aside or bracket off the notion that what is possible is only what is coherent-with-itself. By being indifferent to the form determinations, thought is indifferent to the laws of identity and non-contradiction. To be indifferent is not to destroy or dismiss completely, but merely to withhold temporarily and to see what will come of this. Hegel proposes that it is only by way of this indifference to the form that we become able to explain how to actualize possibility as one whole, to mediate between A and -A, to turn what would seem to be the contradiction of actualizing possibility into affirmative difference. Since the form determinations lead only to (1) the contradiction of actuality that is itself the contrariety of possibility (premise seven); or (2) the reflected actuality of mere possibility (premise eight); or (3) the contingent actuality that remains indifferent to the content of its other (premise ten)—what is necessary is an actuality that can act from within content-differences with others. But this requires not only the existence of the other as what is merely posited alongside and as what could have become actual (as with formal contingency), but also the actuality of the other as contextually integrated within the content of the given actuality. Contrary to formal contingency, the possibility of the other becomes of itself the existing actuality. This is why Hegel turns to "multifarious content" (*mannichfaltiger Inhalt*), to "real actuality" (*reale Wirklichkeit*), and to a contextual world of "diverse determinations" (*verschiedene Bestimmungen*).

Hegel's description at this point in the argument is of a world of conflict, transition, alternation, and becoming-other. While the tree has the possibility of setting its roots in the soil, the mole has the possibility of burrowing a path where the roots would grow. This is no longer a case of thinking self-identity by erasing the possibility of the other at the point of actualization. These conflicts really exist in the soil and in everything that surrounds the action taking place. These conflicts express the contrary nature of possibility, no longer as what must be removed for any actualization to occur, but as the very fabric of the actualization itself.

Hegel claims provocatively that actuality "can act" ("*Was wirklich ist, kann wirken*") (WL 208). Since real actuality emerges from the possibilities that it produces, this actuality is no longer only abstractly identical with itself, but is identical with itself only insofar as it is action.[1] This means that the identity of the actuality comes from the dynamic movements of the engagement of itself in others. Looking ahead at premise seventeen, Hegel will explain this identity as dispersed actuality, where the possibilities of one actuality exist in the actualities of others.

> Die reale Wirklichkeit als solche ist zunächst das Ding von vielen Eigenschaften, die existirende Welt; aber sie ist nicht die Existenz, welche sich in Erscheinung auflöst, sondern als Wirklichkeit ist sie zugleich Ansichseyn und Reflexion-in-sich; sie erhält sich in der Mannichfaltigkeit der blossen Existenz; ihre Aeusserlichkeit ist innerliches Verhalten nur zu sich selbst. Was wirklich ist, kann wirken; seine Wirklichkeit gibt Etwas kund durch das, was es hervorbringt. Sein Verhalten zu anderem ist die Manifestation seiner, weder ein Uebergehen, so bezieht das seyende Etwas sich auf anderes; - noch ein Erscheinen, so ist das Ding nur im Verhältniß zu andern, ist ein Selbstständiges, das aber seine Reflexion-in-sich, seine bestimmte Wesentlichkeit, in einem andern Selbstständigen hat. (WL 208)

> As such, real actuality is, at first, the thing of many properties, the existing world. But this is not the kind of existence that dissolves itself into appearance,

> but, as actuality, is the same as being in-itself and self-reflection. It preserves itself in the multiplicity of mere existence. Its exterior is an inner relation that it has only to itself. What is actual (wirklich) can act (wirken). Things emerge through that which they produce. Their relation to others is the manifestation of themselves. This manifestation is not a transition, not the relation between something and an other in the sphere of being—where the thing is only in relation to another, and although self-subsistent, has its self-reflectedness and its determinate essence in the self-subsistence of another.

When he says that real actuality is on the one hand "the existing world" and on the other hand "the thing of many properties," Hegel exposes two dimensions, one general and one specific.[2] It is difficult not to conflate these two dimensions because each dimension seems to come from the other, and throughout his theory of Real Modality, Hegel often seems to ignore the distinction between them. Still, it is important to distinguish the sense in which real actuality is an actuality of content in general, the whole existing world, from the sense in which actuality is a specific thing endowed with properties that has potentiality, can act, and produce itself.

When he says that real actuality is the existing world, Hegel invokes the exposition from premise one, the starting point of modality, that actuality is immediate existence. However, actuality is no longer immediate existence in this purely formal sense. Actuality is rather the existence of a contextual world. Here, Hegel no doubt makes use of the ambiguity of two connotations of the term "existence": existence is both the immediate being of what simply *is*, and at the same time that which comes into being by satisfying the conditions by which it exists. Actuality is existence, and in a sense it does immediately appear as "merely given" and as "already there"; however, since what appears is a *whole world* of interlocking relations of real limitations, the immediacy of what exists at the same time produces its existence by preserving itself against the manifold of mere existence.

The two operative terms are "to preserve" (*zu erhalten*) and "to produce," literally "to bring forth" (*hervorzubringen*). On the one hand, actuality "preserves itself in the multiplicity of mere existence." On the other hand, "things emerge through that which they produce." These are two process-descriptions for how real actualities relate in an interlocking structure of the contextual world. An actuality preserves itself *against* other actuals, in the sense that it establishes the *real limits* of what it is and what other actuals *are not*. But this preservation is also *preservation for*, in the sense that the actual preserves itself *in the other* as the integration of itself and the other. Likewise, an actuality *produces itself*, in the sense that it is at first only something that might come about, but must become itself *in the possibilities of others*. When something produces itself, it becomes the explicit version of what it already implicitly is.

A brief note on Burbidge's analysis will help to clarify the next transition from real actuality to possibility. Burbidge claims that the reason why possibility consistently comes after actuality in Hegel's argument is because one can only survey what is possible after recognizing what is already actual. This is one of Burbidge's main contributions to the scholarship on the chapter. He lists a formal version of this claim when he explains that for Hegel "possibilities arise from actualities, and not vice versa" (SC 17). This insight gains further traction when we think of how real possibilities only come after real actualities, of what is really possible from the stance of what is already actual. It turns out then that possibilities are not predetermined but are contingent upon the actuality that precedes it. Since the actualities are contingent upon (as per premise eleven), they transform, and since they transform, the avenues that they lead to in possibilities also transform. With the invention of the automobile around the turn of the twentieth century, for example, the roads that were already built for the horse and buggy led to further possibilities, but only from the terms of the automobile, and only after its invention. There is also a certain quality and way of life that is not as possible once the automobile replaces the horse. This, again, is because the actuality precedes what is really possible, because possibilities only appear as what could become of what already is.

15. Real actuality is real possibility.

> Die reale Wirklichkeit hat nun gleichfalls die Möglichkeit unmittelbar an ihr selbst. Sie enthält das Moment des Ansichseyns. (WL 208)

> Diese Möglichkeit als das Ansichseyn der realen Wirklichkeit ist selbst reale Möglichkeit, zunächst das inhaltsvolle Ansichseyn.—Die formelle Möglichkeit ist die Reflexion-in-sich nur als die abstracte Identität, daß Etwas sich in sich nicht widerspreche. Insofern man sich aber auf die Bestimmungen, Umstände, Bedingungen einer Sache einläßt, um daraus ihre Möglichkeit zu erkennen, bleibt man nicht mehr bey der formellen stehen, sondern betrachtet ihre reale Möglichkeit. (WL 208)

> Likewise, real actuality also has possibility immediately present within it. It contains the moment of the in-itself.

> This possibility that is the in-itself of real actuality is itself real possibility, the in-itself as full of content. Formal possibility is self-reflectedness only in the sense of abstract identity, in the sense that something is not self-contradictory. When we bring into account the determinations, circumstances, and conditions of something in order to access its possibility, we are no longer contemplating formal possibility, but are instead considering real possibility.

Hegel restates the argument from premises two through five briefly in terms of real actuality as the contextual world. Just as formal actuality contains possibility (premise two) as the reflection of actuality into itself (premise three), real actuality also contains possibility. But in this case, whereas formal actuality only contains possibility as its abstract identity, by way of positing the equal possibility of the actuality's other, real actuality contains possibility as its concrete identity, as its determinate other. Because real actuality is indifferent to the form determinations, possibility now goes through the mediation of determinateness. This means that possibility as the reflection of actuality into itself becomes the relating ground between something that is determinate against others that are equally determinate. Real possibility becomes the comparing relation between A and B.

-A is the immediate other of A in the sense that something possible can and can not be. But B also signifies the other of A, only not in the abstract terms of self-identity. B is rather the mediated other of A, in the sense that if what is really actual is possible, this is established only from within an interlocking community of determinate others. The example Houlgate gives to distinguish formal from real possibility is particularly helpful at this stage of the argument: "It may well be possible in itself for me to be the tallest person in the world; however, it is clearly not possible for me to be the tallest person, to the extent that there are, and always will be, others who are taller than I" (NC 40). Houlgate's example would make more sense if he had indicated his actual height and explained that others in this contextual world are indeed taller than he is. Nevertheless, his example helps to clarify this distinction between the initial formal possibility, the self-identical reflection of something into itself, the abstract possibility that Houlgate could be the tallest person in the world, and the real possibility, the possible relative to others in this real, contextual world, where Houlgate is taller than some but shorter than others.

Hegel says that we observe this exposure of the content between things when we "bring into account the determinations, circumstance, and conditions of something." To posit the real actuality of something is to posit along with this a whole field of "determinations," "circumstances," and "conditions." These structure something's possibilities in the contextual world. They reveal that something is only itself if it is understood in the terms of its possibilities with others. We can already anticipate Hegel's next step, that if something is only understood in its relations with others, it is only itself insofar as it is these others, which it is not.

In the quoted passage for this premise, Hegel does not precisely define these three structures, which each explore how real actuality entails itself in the possibilities of others. On my reading, if one brings into account "the determinations," this means to recognize the specific limitations between something and its other. But if one brings into account "the circumstances," this means that real actuality is an existing multiplicity that contains many actuals within it. Yet if one brings into account "the conditions," this means to describe how something becomes itself through possibilities that are dispersed in others. I realize that Hegel might not have intended to separate these three terms as I have just defined them. He might have listed them in a more off-handed way, and he might think of all three at once as his explanation for the entailment from real actuality to real possibility. I also realize that Hegel seems to want to move directly to "conditions," "dispersions," and "actions-on" and that it might be a delay of his argument to investigate what he means by "determination" and "circumstance." But I would still like to linger for a moment on this because I think the short detour adds important details, which Hegel himself should have explored.

"Determination" (Bestimmung) is the term that Miller translates as "limitation" from the very beginning of the *Logic*.[3] The limit dictates the content, or, in Hegel's initial terminology, the "determinate being" (Dasein), between something and its other. We think of actual determinations as separate from each other and as having a limit. But insofar as each is the limit of an other, each is not only itself but a going-over-into the other. Here Hegel's claim that possibility ought-to-be the totality takes on a further dimension, in the sense that if something ought-to-be, this posits both the limit that something is not its other, but also the trajectory that something should be both itself and its other. If something ought to be its other, then it is both not its other, but is also a going-over-into its other.[4] Limit is that which is shared between that which is different. The limit both distinguishes each from its other, but at the same time makes the content of each dependent upon the differences of the other's content. For example, the color blue can be distinguished on the color wheel from the color green not only because it is an inherent property, but also because blue is not green. The wheel itself draws the limit between each color. Without this "drawing up" of limits, each color would dissolve into the others and become "grey." It is the limit, then, and not only the inherent quality of the "this," that produces the content-differences between each. Moreover, since each does not possess its own limit, but shares this limit with its other, each "this" is not only reversible (i.e., something is the other of another something), but really exists in the other's content. This means that the something exists as its other inasmuch as it exists as itself. The structure of "determination" will become more apparent at premise sixteen, where Hegel discusses the dispersion of actuality.

"Circumstance" (Umstand) is the term for existing multiplicity. On my reading, "determination" and "circumstance" describe a similar structure, only from different perspectives. While determination describes something's relation to its other through limit, circumstance describes the contextual "world" surrounding the determination of each. Circumstance describes not only the limit between properties, e.g., between blue and green, but also between a multiplicity of background situations that all things share with each other. If we are asking about the circumstance behind something, then we are asking about the situation that sets the something into its specific position and produces its place in the community. Circumstances might include specific collections, events, coordinates, genealogies, taxonomies, histories, etc. The linguist Ferdinand de Saussure (1857–1913) makes a case for this structure in terms of language. De Saussure argues that language is "a system of interdependent terms in which the value of each term results solely from the simultaneous presence of the others."[5] A critique of the assumption that the meaning of each word is contained within the boundaries of that word alone in isolation from others, de Saussure argues that meaning comes from the community of the words that surround it and that together form the circumstance of language. Each word holds the others in place, just as the others hold it and produce a place for it. "Within the same language," de Saussure says, "all words used to express related ideas limit each other reciprocally; synonyms like French redouter 'dread,' craindre 'fear,' and avoir peur 'be afraid' have value only through their opposition: if redouter did not exist, all its content would go to its competitors."[6] He might have gone too far in claiming that the value of each is only in the opposition of others, or that if one did not exist, then all its content would go to the others. Even calling the others "competitors" might sound a bit too extreme for Hegel's position. Still, there is something in the sentiment of de Saussure's claim that helps to expose what Hegel means by "circumstance": a co-dependency of each from the stance of others.

Finally, "condition" (Bedingung) is the term for actualities that are at the same time the possibilities of other actuals. Water is a condition for the seed becoming a plant. Of course water is an actuality as well, but it is also a possibility that creates a relay between something that is initial and something that results from this. While determination is about the limit that something shares with its other and circumstance is about how the context produces a place for others, the condition describes this same point from the disposition of something that becomes itself in external possibilities that are at first literally in the actuality of others. How something comes into itself as its other via conditions is, arguably, the most significant component of Hegel's argument from real modality.

To posit the real actuality of something is to posit along with this a whole field of determinations, circumstances, and conditions. These are the structures of possibility in the contextual world. They reveal that something is only itself if it is understood in the terms of its possibilities with others.[7] We can already anticipate Hegel's next step: if something exists only in its relations with others, it is only itself insofar as it is these others, which it is not.

16. Real possibility is an existing multiplicity.

> Diese reale Möglichkeit is selbst unmittelbare Existenz, nicht mehr aber darum, weil die Möglichkeit als solche, als formelles Moment, unmittelbar ihr Gegentheil, eine nicht reflectirte Wirklichkeit ist; sondern weil sie reale Möglichkeit ist, hat sie sogleich diese Bestimmung an ihr selbst. Die reale Möglichkeit einer Sache ist daher die daseyende Mannichfaltigkeit von Umständen, die sich auf sie beziehen. (WL 208–9)

> This real possibility is itself immediate existence, but no longer because the possibility as such, in its formal moment, is immediately its opposite, an unreflected actuality. Instead, because it is real possibility, it has this determination directly within itself. The real possibility of a thing is, therefore, the existing multiplicity of the circumstances, with which it stands connected.

We have already discovered in formal terms the limitless multiplicity that everything is possible as long as there is no contradiction. This is the case because, just as possibility presents the positive identity of actuality (that if A is possible, then A is A), likewise it presents actuality with the opposite of itself. This same process now happens in terms of real possibility. But now the process happens, not just between the actuality and its abstract opposite, its non-actuality, but between determinate actuality and other determinate actuals. Since what is really possible opens actuality to the determinations, circumstances, and conditions of itself, the relationship between actuality and possibility becomes one of existing multiplicity. Marcuse gives a vivid example of existing multiplicity when he writes,

> The upright tree in the forest can be hit by lightning, can collapse, can dry up, can be sawn as wood and utilized as construction material. All these possibilities belong to the in-itselfness of the tree; its actuality can pass through all of them. All these possibilities are themselves always already actually there: the electrically charged atmosphere, the woodcutters, the sawing mill, the building to which the planks will be transported, all exist somewhere. When they actually become possibilities of the tree, this plurality of possibilities also becomes an "existing multiplicity of circumstances" which "relate" themselves to the tree. The tree undergoes all its possibilities as actuality. It moves itself through them as "the same" tree. (HO 94)

In Marcuse's example, the determinate possibilities of the tree (that it can be hit by lightning, collapse, dry up, be sawn, be utilized) constitute the movements and transformations of the same actuality throughout. But on the other hand, these determinate possibilities carry the actuality beyond itself. This is the case because the real possibilities of something are both what allow this actuality to act, but also what determine the limitations that stand against this particular actuality, disseminating its content and defining the parameters of its finitude. The tree cannot be anything and everything but must follow the path of the limitations and determinations. As we will see at premise twenty-one, the limitations of something's determinateness are at the same time its possibilities. Real actuality can act because the possibilities that it entails are not something abstractly opposite to what it is, as in the sheer other of formal possibility, but are now the determinate possibilities of what it can become insofar as it can produce itself.

Whereas formal multiplicity is limitless in the sense that possibility is as open to the non-being of its actuality as it is to its being, existing multiplicity is rather just those specific circumstances that are connected to something and make it possible. This real type of multiplicity, although it is certainly not limitless since it is in a sense generated from limitations, is nevertheless a multiplicity that maintains the shape of the actuality's negation. Existing multiplicity is the limited diversity of the circumstances that surround something determinate. It is also the site where the action takes place, when the actual acts upon itself as upon an other.

If real actuality contains real possibility as an existing multiplicity, the consequence is that the possibilities of something are not simply its own. Since its possibilities exist in the circumstances that surround it, something is really itself only insofar as its content is simultaneously the content of others. The circumstances that surround what something is constitute an interlocking network of multiplicity. Many dispersed actualities, held together in possibility, constitute one "large" actuality of multiplicity. This large actuality presents the whole situation, the entirety of the circumstance, the complete interlocking network of relations between actuality and possibility. In this sense, real actuality is both a determinate thing that has properties, can act, and produce itself, but it is also an actuality *of* multiplicity, an actuality that presents the whole range of determinate possibilities, whose content is the content of others.

Remark: Heidegger's Greek Temple and Existing Multiplicity

Heidegger's description of the Greek temple in *The Origin of the Work of Art* presents us with an example of what Hegel means by existing multiplicity. The temple sets up a contextual world of real possibilities and dispersed actualities.[8] Heidegger writes,

> [The temple] . . . first fits together and at the same time gathers around itself the unity of those paths and relations in which birth and death, disaster and blessing, victory and disgrace, endurance and decline acquire the shape of destiny for human being. . . . [The temple] holds its ground against the storm raging above it and so first makes the storm itself manifest in its violence. The luster and gleam of the stone, though itself apparently glowing only by the grace of the sun, yet first brings to the light of the day, the breadth of the sky,

> the darkness of the night. The temple's firm towering makes visible the invisible space of air. The steadfastness of the work contrasts with the surge of the surf, and its own repose brings out the raging of the sea. Tree and grass, eagle and bull, snake and cricket first enter into their distinctive shapes and thus come to appear as what they are. . . . The temple, in its standing there, first gives to things their look and to men their outlook on themselves.[9]

The temple is not merely a building inserted into the space of Greek society. It is not merely a thing among things that exists on its own and can be placed and replaced without transforming its internal content. The temple is the site of multiplicity. It sets up a contextual world. The storm that strikes above it appears because of its relationality with the temple. The stone carries with it the meaning of day and night, of light and darkness. Heidegger even claims that the different types of animals—"the eagle and the bull, the snake and the cricket"—gain their distinctive shapes through the context of the temple.

Although Heidegger focuses primarily on the temple as what sets up the world, we can assume that the temple is equally grounded in the world of Ancient Greece. Just as the temple sets up a world for the animals, shapes the destiny of humans, and brings context to the historical people of Greece, the Greek world and everything that surrounds the temple, from the plants and the animals to the sun and the rocks, all frame the temple as one actuality of existing multiplicity.

We see this point more clearly when we think of what it would mean to take the temple out of its context in the Greek landscape. What would happen, for example, if the Greek temple were "torn out of [its] own native sphere"[10] and transported to the Metropolitan Museum in Manhattan? The temple would not carry the Ancient world to Midtown, as one might assume. When transplanted to Manhattan, it becomes displaced and no longer is what it is. Of course, when surrounded by the circumstances of the Metropolitan Museum, Central Park, and Midtown Manhattan, the Greek temple becomes another sort of thing. Its inner content becomes "shaped" within the circumstances of the museum. One important element of these circumstances is that when the stone blocks are reset piece by piece within the cold air-conditioned room of the museum, they still resemble in their displacement the temple as it once was, set on the hill in Greece. This displacement and replacement becomes part of the world that the museum opens for the destiny of humans.[11]

17. Because of existing multiplicity, possibilities are dispersed in the actuality of others. The consequence is that something's possibilities are not its own but are always deferred to others.

> Diese Wirklichkeit, welche die Möglichkeit einer Sache ausmacht, ist daher nicht ihre eigene Möglichkeit, sondern das Ansichseyn eines andern Wirkli-

> chen; sie selbst ist die Wirklichkeit, die aufgehoben werden soll, die Möglichkeit als nur Möglichkeit.—So macht die reale Möglichkeit das Ganze von Bedingungen aus, eine nicht in sich reflectirte, zerstreute Wirklichkeit, welche aber bestimmt ist, das Ansichseyn aber eines andern zu seyn und in sich zurükgehen zu sollen. (WL 209)

> This actuality, which constitutes the possibility of the thing, is therefore not its own possibility, but the in-itself of another actuality. It is itself the actuality, but the actuality as what should be sublated, the possibility as only possibility (*qua* possibility). This is why real possibility constitutes the totality of conditions, a dispersed actuality, which is not self-reflected, but is determined, instead, to be the in-itself, but of another, and is supposed to have returned to itself.[12]

Hegel describes existing multiplicity as a process of dispersion. The possibilities of something are not simply its own but are rather the possibilities of other actuals. In this respect, dispersed actuality offers a significant advancement upon the earlier disposition of formal contingency. Formal contingency had presented the sides of possibility indifferently, by positing the equal existence of the non-actual other. This expressed the possible itself, but in an inadequate way, because contingency still held the contrary sides of possibility apart from itself. Dispersed actuality fixes this problem by dispersing possibility into the existing actuality of others.

If the possibilities of something were simply its own, and not dispersed in others, this would return us to the contradiction at premise seven. The possibility of the "can not" would expose the actuality in an immediate way to the opposite of itself *as itself*. Since this does not work, actualization goes through the *mediation* of possibles in other actuals, which, when taken together as one process, should complete all possibilities in an actuality of multiplicity.

However, dispersed actuality presents us with a further complication. Since each actuality defers to the actuality of others, possibility does not seem to reside in any one of these actuals, and therefore does not seem to maintain its own existence independently of others. One of the main points Lampert makes in his essay on contingency is to explain this problematic of dispersion. We often think of possibility as an attribute of actuality, but Lampert claims that the possibility of an actuality, for Hegel, is the possibility of another. If each actuality is the possibility of other actuals, then "the truth of one thing is *in* a different thing" (FM 77). But this conclusion leaves actuality without an accountability of its own possibilities. If each actual can find itself only in the relay of its possibilities in other actuals, each actual is like a hollow shell, signaling only that one would need to look elsewhere to realize what it itself is. On the one hand, possibilities would slide around in a vicious circle of deferral. On the other hand, actualities would be unable to give a stable account of themselves because they are only the possibility of others.

Hegel then turns to conditions because conditions explain why the process of dispersing one's possibilities in other actuals is really the process of actualizing one's own possibilities.

Remark: The Relativity of Active and Passive Possibility

When one takes into account the determinations of something, real actuality has real limitations and stands against others, which equally limit and determine what something is. We can describe this as a relation of external comparison. The cat is relatively big when compared with the mouse. The mouse is relatively small when compared with the cat. Each is determinate against the other because each is itself alone and not the other. Yet, each is determinate because of the other, in the sense that the qualifications big and small fill the cat and mouse with content only if each is measured against the other. Compared to an adult bear, the cat is quite small. The cat has claws and these are dangerous, but only against that which can be maimed by them. The claws are dangerous to the mouse but merely annoying to the bear. The external comparison distributes the content to each. As the comparison changes, the content changes as well, even to the extreme point where, through one comparison, something has the opposite content from what it might have had were it compared differently.

But the truth of this external comparison is in its underlying immanence. The relativity reveals how the possibility of something exists in the possibility of others. The comparison is in this sense immanent in the other, rather than located in the external relativity of the relation. If you want to know about the cat's claws, look at the mouse or the bear because the strength of the claws really exists in them.

This amounts to a modal insight about the correlation between active and passive possibility. Something only has the power to act upon another, or upon itself as if it were an other, if that which it acts upon has a passive but reciprocal power to receive it. Aristotle makes this point in Book *Theta* of the *Metaphysics* (CWA 1046a19–1046a26 and 1021a15–1021a19). Fire can only burn that which can be burnt. It can burn paper because paper has the passive power to be burnt; it cannot burn a rock since a rock does not have the ability to receive the fire. The fire's ability to burn the paper is in this sense dispersed into the paper's ability to be burnt.

The passive possibility to receive is the condition by which something active can act. But if it is the condition for the possibility of action, then the passive is truly active. Hegel calls this solicitude (SL 521–23). The passivity of the other actively solicits the activity of something upon it. In other words, if something solicits, it determines what is active to act upon it. This turns the passivity on its head.

In the "Mechanism" chapter of the *Logic*, Hegel adds that sometimes a determinate thing will have to prepare the other to receive it. Sometimes the passivity of the other is too weak to receive the activity. Such weakness effectively disarms the active agent. To become determinate, the thing must draw its other up to its dimension. Hegel gives a strange yet vivid example. If a bullet from a gun is to tear through a cloth sheet that is hanging on a line and fluttering in the wind, the cloth sheet must be made rigid by applying a shellac so that it becomes taut. The cloth must be prepared so that it can stand up against the bullet and receive it. This example contributes to Hegel's point that there is a correlation between active and passive possibility, and that this correlation is a dispersion of the possibilities of an actual into other actuals. "If they were not in the same sphere," Hegel says generally about objects in relation, "their relation to one another would be an infinite judgement, and no process between them would be possible" (SL 719).

Remark: Possibility Transference

We can also interpret Hegel's use of the term dispersed (*zerstreute*) quite literally. If the possibilities of something are the possibilities of other actuals, then things give away that which makes them determinate. Other actuals carry out the possibilities of something. Now, this can mean that the way something accesses its possibilities is by the use of another, as with the master-slave relationship in the "Lordship and Bondage" passages of the *Phenomenology* (PS 111–19); or this can mean that something places the legacy of its possibilities in another actual, as when a parent transfers possibilities to a child.

In the case of "Lordship and Bondage," the slave carries out the master's desires. The master channels her possibilities through the slave and forces the slave to shape the world for her. But since it is the slave who works to shape the world, it is really the slave who has possibilities and can act. The master has her possibilities dispersed into another actual, the slave, and it is through this relationship of coercion that the master is able to satisfy her desires. This means that the master's possibilities are not her own, but only the possibilities of the slave. The analogy, especially in terms of the reversal of master and slave, shows that although it would seem that the something should have the possibilities, these possibilities are truly the possibilities of another actual. This means that the something does not have its own possibilities but is merely what spurs the activity.

Lampert takes this point further when he claims that possibility can be transferred between generations, as in the case where "having children is passed on to one's children" (FM 75). The transference between generations happens in two senses. The actuality of the parent has at least a portion of her possibilities in the actuality of her child. The legacy of the parent is acted out in the growing maturity of the child. The parent educates the child by transferring knowledge, habits, and customs. The parent hopes that she will live on in the actuality of her child, that what she has developed of herself will carry over into the other, who will mature and live beyond her. But the transference also happens, as Lampert emphasizes, in the sense that the potential to generate offspring is passed on to the next generation.

In certain other cases, by giving over one's possibilities to another actual, something can gain access to entire regions of determinate content that would otherwise remain inaccessible. For example, Hobbes argues in *Leviathan* that to enter into civil society requires that we transfer to the authority of a sovereign the natural ability that each of us has to harm the other. By this transfer, each of us gains the ability to trust that others will not infringe upon our basic rights even though this is possible. By transferring possibilities in respect to another, there are other possibilities that become transferred to you. In the case of Hobbes's social contract, the power you receive, that by law the other must not infringe upon your basic rights, is not a natural ability, but an ability that you can access by giving up your own ability to infringe upon others. The contract comes about, in this sense, as a possibility transference.

18. Since the possibilities of dispersed actuality do not seem to reside anywhere at all, something becomes actual through possibilities that are dispersed in its conditions.

> Wenn alle Bedingungen einer Sache vollständig vorhanden sind, so tritt sie in Wirklichkeit;—ie Vollständigkeit der Bedingungen ist die Totalität als am Inhalte, und die Sache selbst ist dieser Inhalt bestimmt eben so ein Wirkliches als Mögliches zu seyn. (WL 210)

> When all the conditions of a thing are completely present, it enters into actuality; the completeness of the conditions is the totality as in the content, and *the something itself* is this content determined as being equally actual and possible.[13]

Actuality is dispersed, but not only in the sense of determination, where the actuality of something has its possibilities in other actuals. While this description of dispersion is not incorrect, it leads to the conclusion that one's possibilities are not really one's own, but only the possibilities of others. To expose this subtlety, Hegel turns to dispersed actuality in the sense of conditions, where something initial becomes actual through possibilities that are dispersed in others. That actuality is dispersed suggests, as one whole, that it has become divided and compartmentalized by the plurality of possibility. It now carries the properties of all its others as its own property. Hegel turns to conditions to explain how the possibilities in others are still one's own.

Many commentators mention the role that conditions play in the "Actuality" chapter; however, no one examines this concept in great detail. Lampert discusses how dispersion, transference, production, and various other movement-structures work in terms of multiplicity, and analyzes the function of conditions at the same time (FM 77). His analysis directs us to think about conditions as movement and becoming. Houlgate says that the reason why immediate actuality is contingency is because it harbors within this concept "the possibility of something else arising" (NC 43), that is, the contingent actual as a condition. Burbidge presents the argument that conditions are always multiple and that no one condition can exist alone (SC 34–37). This is a good insight on its own because it explains why conditions are always a multiplicity, but Burbidge does not thereby fully emphasize why a condition is a possibility and an actuality together. Ng claims that if Hegel's conditions of possibility are full of content (not empty and formal), the consequence is that there can be no real distinction between the empirical and the *a priori*. Ng's remark explores the question of whether Hegel is playing on two senses of conditions, the first, the Kantian question, what are the *a priori* conditions of possibility for any experience whatsoever?—the second—a more literal usage, the material conditions of possibility for something's result in actuality.[14] On my reading, Hegel's claim about conditions is one of the most important and controversial premises of his entire argument because this establishes a unity that is as much possibility as it is actuality, thereby explaining how possibility and actuality are transitional concepts.

Conditions are both actuality and possibility. They are actual in the sense that they are immediately given as the fact of existence. For example, insofar as the stone is a condition for something, it is immediately actual and first appears as the earth itself. But conditions are also possibility in the sense that they have latent within them the result of further actuality. The stone is given as prior, but it carries in its immediate content the further possibilities of the statue, the house, the street, and so on. Conditions are actualities that are not just themselves. In *the Encyclopaedia Logic*, Hegel claims that they get used up (*verbraucht werden*) or sacrificed (*aufopfern*) in the process of actualization (EL 220). They are immediate actualities but also the material for what can become actual if the conditions themselves are completed in the process of actualization.

The possibility of something stands latent in the immediate actuality of the conditions. The actuality contains the possibility as latent within it, and this latency itself is the compulsion of the further entailment from possibility to actuality. One only needs to draw out the possibility from the actuality, and the further actuality will result from this.

However, this also means that the possibility nevertheless completes itself only in the resulting actuality. It would seem that not any possibility stands latent within the conditions, but only certain possibilities entailing certain actualities. The stone has its end in the resulting statue. It can also be used up in the actuality of the house or in the cobble of the street. But it would seem that the stone does not have within it the latent possibility of significantly different, non-teleological actualizations (of becoming an elephant or a kitchen window, for example). The entailment from possibility to actuality would seem only to go in certain directions. It is worth noticing now that one of the main characteristics of absolute modality, which will come from the exposition in chapter 3, is to establish a case for why conditional actualities contain every possibility within them, not only certain possibilities entailing certain actualities.

What is important to highlight from this moment of the argument is that real actuality contains real possibility as the *latency* of further actuality. What immediately appears as the fact of existence carries within it the possibilities of others. But these others are not really "other" from what immediately appears. The stone, after all, does not become something other than the stone when it takes the form of a statue, but remains this same material throughout. Insofar as possibility is latent in the conditions, it is the comparing relation, no longer between what is immediately actual and the abstract opposite of this, but rather between what something initially is and what it could become.

We can read in the connotations of the term *possibility-latency* the same compulsion that Hegel has already established at premise thirteen in terms of formal necessity. If what is actual is possible (premise two), then actuality *cannot* only be one side or the other of the possible, but must come to form itself as the totality of possibility. The other, although it would seem to remain against the actual as what the actual is not, must become consumed in the process of actualization. This is why, although conditions are immediately actual, they are also the possibility of something initial coming to its end in actuality. Conditions "fall under" as actuality becomes itself. Although one's possibilities are dispersed in the externality of other actuals, since these "actuals" are only the conditions for the possibility of something initial coming to its end in actuality, the thing in question nevertheless remains self-coherent throughout this process. By showing how something initial results in actuality through the conditions of others, Hegel establishes the argument that the possibilities dispersed in other actuals nevertheless remain the something's own possibilities throughout.

Remark: How to Make All Conditions Completely Present

In paragraph 148 of *the Encyclopaedia Logic*, Hegel divides conditions into three moments:[15] (1) material conditions, (2) the thing in question, and (3) activity (EL 224). The conditions (*die Bedingung*) are the material requirements, the immediate actualities, that get used up in the process of the actualization. These actualities begin as external limitations that stand against the actualization, but if they are engaged or consumed, they help to constitute the materiality of the developed actuality. However, we should also recognize that there are various senses of condition. Hegel tends to focus on conditions as that which get used up and consumed, but there are also conditions that can be used but not used up, as in the case of a hammer, which when used for mounting a picture on the wall, does not become used up in the way that the nail disappears when it is hammered into the wall.[16] The thing in question (*die Sache*) is the initial possibility of what could become actuality. It represents the entire process that there is something initial which has the possibility, if it satisfies certain conditions, of coming forth into actuality. And third, activity (*die Thätigkeit*) is the active agent that uses or uses up the passive conditions at the point of material actualization. There is also in a sense a fourth moment, which Hegel describes in paragraph 147 as the developed actuality (*die entwickelte Wirklichkeit*) (EL 220–21). This is the final result, but also the complete reality, of the thing in question.

Hegel sometimes describes conditions as "material" but Béatrice Longuenesse suggests that conditions can also be "spiritual." By "spiritual" she means historical, economic, social, geographic, and climatic conditions (CM 135). Hegel probably does not mean to exclude these strictly non-material versions of conditions. By "conditions" Hegel includes anything necessary for the mediated actualization of something initial into something actual, whether this is strictly "material" (e.g., stone, wood, blood, etc.) or "spiritual" (e.g., the disposition of mutual respect, the conditions for an economic boon, the deed to build a house, the license to drive a car, etc.). Prior to resolving themselves in actuality, each condition is similar to unformed (or semi-formed) matter, in the sense that unformed matter holds the possibilities of formed matter within it. Certainly, each condition is an immediate actuality on its own, and in this sense already contains order and form (e.g., the stone itself is form, not just the matter of the house), but in as much as each condition is the possibility of another, each appears first as materiality and then becomes the formation of others as the result of actuality.[17]

Since the passage quoted with premise eighteen is quite intricate, let us analyze the mechanics behind this process of actualization, which Hegel describes as happening only when "all of its conditions are completely present" (*Wenn alle Bedingungen einer Sache vollständig vorhanden sind*).

A. What does Hegel mean by "completely" when he says that all of the conditions become completely present?

If there are many conditions and each is completeable, this suggests that they are finite but also that they belong in various sets. They are finite in the sense that they are exhaustible, in the sense that they get used or used up in the process of actualization. But they are also finite because they belong to one another, as a member belongs to an exclusive set. While the actuality of the house is latent in the possibility of the stone, the stone itself is insufficient to entail its end in the house. This actualization requires a whole set of other conditions: other material such as wood, nails, brick, plastic, cement; but also active agents such as builders, plumbers, carpenters; and even immaterial social conditions, such as a permit from the city, a loan from the bank, a deed to the land, and so on.

In the same passage, Hegel calls real possibility the circle of conditions. "Because, as the circle of conditions, the immediate existence of real possibility sublates itself, it makes itself into the same in-itself that it already is, in other words, it makes itself into the in-itself of an other" (WL 210). Various conditions come together to form a "circle." The completion of all conditions together in one "presence" is what it would take to release the possibility from its containment in the immediacy of each actuality.

This process of gathering together the various conditions of something is made more complicated by the fact that each condition contains within it a multiplicity of other possibilities, which could in turn result in actuality. The stone itself has a seemingly indeterminate amount of possibilities latent within it. Certainly, the stone house is a possibility in the stone, but the statue, the street, the arrowhead, and innumerable other developments also exist within it, and only need to be drawn out. Since these possibilities often stand in opposition to one another (for example, the dirt of the earth is a condition for the roots of the tree but also for the mole's burrow), Hegel returns again to the problematic of premise six, where the limitless multiplicity that everything and anything is possible leads to diversity, which leads to opposition, then to contradiction.

B. If all conditions become completely present, does this mean that they must become present, or only that actuality results from them if they all happen to be present?

Is Hegel saying that the process of actualization must find all of the hidden and obscure corners of an actualizable set in order to bring out the possibility? Or is he merely saying that all the conditions would need to become completely present for the possibility to result in actuality? If he means the former as a prescription, then the obscure and the hidden would be of great importance, since only by uncovering what is obscure of the range could the process of actualization come to complete itself in actuality.

This is why designers of board games test the structure of the game before they release it to the public. They look for the most obscure and unusual variations for how one might play the game. Only by coming to terms with the abnormal inconsistencies that a given set of rules might have within it do the designers come to realize exactly what the game is. They can then adjust and change what it is in order to strengthen its design. We can conclude from this example that there is something important about uncovering all conditions, and not only those which are the most prominent and obvious for a given set. But whether this is always the case or not remains difficult to answer from what Hegel has written.

C. What does it mean to make all conditions "present"?

Hegel says that something initial becomes actual when all of the conditions are completely present. We can think of this in terms of parts and wholes. The whole appears as complete only if all of its parts are assembled in the right order. If the parts are not present or do not connect in the right ways, the whole remains incomplete, and the thing in question does not become actual. Presence (*vorhanden sein*) is Hegel's term for a process where strands of determinate possibilities come together into a formation prior to their actualization. By making all conditions completely present, this turns the initial something into the resulting actuality.

For example, if all of the conditions of an internal combustion engine are completely present, this means that the engine has all of its integral parts, and these parts all work together to produce the function of the engine. Again, this is why Hegel says that what is really actual can act. The engine burns gasoline and this sets the car in motion. But the engine can only act if the crankshaft turns the pistons back and forth from within the cylinder, if the sump surrounds the crankshaft, if the valves let in the fuel and release the exhaust, and if the spark plugs ignite the air and make the combustion combust. Since these parts are integral, if they are absent, the engine will not function, or at any rate, will not function properly. While there is a difference between not working properly and not working at all, we can assume that if the engine is not identical to itself in parts, then it is not really an engine but only a heap of disassembled machine, which does not result in actuality because it cannot act.

This process is made more complicated by the fact that engineers sometimes build into the structure an amount of redundancy that keeps the engine functioning even when the parts break down or go missing. Sometimes parts are also designed to replace or regenerate themselves. All of this complicates, but does not obscure, Hegel's main point about "presence," that the presence of specific determinate conditions is integral to the process of material actualization.

There are at least two ways in which the intricacy of parts can lead to sustained positions of unactualized possibilities. The first is in the sense of resistance. The more intricate the parts, or the larger and more obscure the range, or the more the parts need to be just so—the harder it becomes to trigger the actualization. The second is in the sense of collaboration of function. The more intricate the parts, the greater the chance that one part will take over while others remain underutilized. In this sense of underutilization, the parts remain untapped and retain the status of unactualized possibilities.

While there might be a debate about what constitutes the integral parts of any given real actuality, and while there might be a lot of options for replacements of conditions, substitutions of order, alternatives of design, and so on, Hegel's point, nevertheless, is that whatever is integral must be present; otherwise, the thing in question will not find its end in actuality. But this also means that if the thing in question were to become overfull by the presence of what does not belong within the circle of its finite set, then this thing would either become absent from itself, dismembered and disorganized, as something falls into chaos, as the shape of form falls into matter, or another determinate thing would rise forth in its place, take hold of the conditions, and turn itself into the actual instead of what was originally proposed.

D. Usually, one enters a place or enters into a conversation. What does it mean for something to enter actuality (so tritt sie in Wirklichkeit)?

If something only enters into actuality by gathering together the parts of itself, the strands of its possibility, one might assume that actuality-entrance is a gradual process. However gradual and piecemeal this process might seem to be, Hegel suggests that at the point of completion, the something enters into actuality *all at once*. In other words, we might use the logic of part and whole when it comes to the gradual completion of conditions; however, this logic is no longer appropriate when we talk about the exact point when something enters actuality. The word *to enter* suggests that something goes through this transformation in its entirety, that the transformation from the conditions of possibility into actuality is immediate and indisolvable.[18] Again, we find ourselves turning almost prematurely with Hegel to the disposition of absolute actuality, where, not one and then another possible condition comes to form the resulting actual in a *successive* way, but rather the whole circuit of possible conditions *co-exist* together as one *simultaneous* actuality. This will become more clear after we discuss Hegel's theory of relative necessity and how contingency stands at the base of this.

19. What is initial can only become actual if it does not contradict the conditions that make it possible.

> Was real möglich ist, ist also nach seinem Ansichseyn, ein formelles identisches, das nach seiner einfachen Inhaltsbestimmung sich nicht widerspricht; aber auch nach seinen entwickelten und unterschiedenen Umständen und allem, womit es im Zusammenhange steht, muß es als das mit sich identische sich nicht widersprechen. (WL 209)

> In accordance with the in-itself, possibility is a formal identity, whose inner content is non-contradictory. But, as self-identical, it must also not be self-contradictory in its developed and distinct circumstances and in everything with which it stands connected.

Hegel returns to his earlier definition of possibility as whatever does not formally contradict itself. But now he claims that real possibility must also follow the principle of non-contradiction and self-coherence. However, the restriction on real possibility is more severe than what remains simply self-identical, because now one must not contradict the determinations, circumstances, and conditions that constitute what something is. Actualization becomes restricted not only to the possible as the non-contradictory, but also to whole regions of content that are not possible from the stance of the thing's contextual environment.

There are at least two reasons why Hegel's claim at premise nineteen might seem problematic: (1) it leaves us to wonder why he would choose to retain the term contradiction instead of simply calling this an impasse, and (2) it leaves us to wonder why, if an actualization were to occur without meeting its basic conditions, this would be a real contradiction and not a logical contradiction. As for the first of these, the question is whether there is enough opposition and negativity in real possibility to call an actualization that fails to meet its conditions a contradiction. Although he can be interpreted to overemphasize the role that contradiction plays at this point in the argument, we can also see why Hegel would want to retain the strength of real possibility's negativity, since real contradiction produces movement.

As for the second of these, although he continues to recognize logical contradiction at this point in the argument, Hegel also wants to establish that there is a further compounding of real contradiction on top of formal contradiction. Formal contradiction presents us with the minimum limitation that something is impossible if it is both itself and the opposite of itself. But by establishing real contradiction on top of logical contradiction, Hegel is able to explore a more detailed mechanics for how to think of something as both itself and the other of itself. It is this compounding of the two contradictions that lets us think about how something initial becomes actual by making all of its conditions completely present. This process of material actualization is the process by which something comes to adapt to the distinct circumstances with which it stands connected, on the one hand, turning itself into something that can accept the conditions, on the other hand, reshaping the conditions to make the parts harmonious at the point of actualization. While everything and anything is formally possible, so long as there is no logical contradiction, what is really possible contains the further limitation that it must remain self-coherent within the environment that determines it.

We might conclude from this premise that if it must submit to *both* the formal and to the real principle of non-contradiction, then what is really possible is a more restrictive and narrow set than what is formally possible. But Hegel will argue, contrary to this, that real necessity is not only a double-restriction (in the sense that something cannot be the opposite of itself, but also must not contradict the context that makes it determinate), but that this restriction itself leads to the only structural relation between actuality and possibility whereby something becomes itself in its other and becomes the other of itself. In this sense, real modality is more inclusive than formal modality because by submitting to real necessity, things become able to maintain what is otherwise formally impossible, that they are both themselves and their others. To see how this works, we will first need to understand why the resulting actuality that comes from the conditions of other actuals is in a different sense from premise nineteen the demonstration of contradiction, and why it is the case that this demonstration leads to movement.

20. But since each condition contains a multiplicity of other actuals, to become actual is to become in contradiction.

> Aber zweytens weil es in sich mannichfaltig und mit anderem in mannichfaltigem Zusammenhange ist, die Verschiedenheit aber an sich selbst in Entgegensetzung übergeht, ist es ein widersprechendes. Wenn von einer Möglichkeit die Rede ist und deren Widerspruch aufgezeigt werden soll, so hat man sich nur an die Mannichfaltigkeit, die sie als Inhalt oder als ihre bedingende Existenz enthält, zu halten; woraus sich leicht ihr Widerspruch auffinden läßt. (WL 209)

> But, second, because it is in a multiplicity with itself and exhibits a multiplicity in relation to others, and because diversity passes over on its own into opposition, real possibility is contradiction. If we want to demonstrate this contradiction, we need only to fixate on the multiplicity that possibility contains as content or as its conditioned existence, and from this the contradiction is easily exposed.

Hegel uncovers the paradox from premise seven now in terms of real possibility. Anything is formally possible that does not contradict itself. But if we take the whole disposition of possibility as one actuality, everything is equally impossible and self-contradictory. The same problem now recurs when taking real possibilities into account. The context that surrounds something and makes it determinate contains diversity, diversity contains opposition, and opposition contains contradiction. Hegel proposes that if we shift our disposition to the multiplicity of the contextual world, it is easy to discover real contradiction.

Let us look at Hegel's claims about real contradiction in more detail. It is not obvious from the text whether Hegel means (1) thought can demonstrate the contradiction by fixating onto the multiplicity *latent within each condition*, or (2) the contradiction appears in the transformation when something initial results in actuality. Hegel probably means both of these at once, as his distinction between "multiplicity with itself" and "multiplicity in relation to others" suggests.

(1) Each condition has a multiplicity of possibilities existing latent within it, but it would be a contradiction to actualize everything that a condition could become. While the log of wood has all sorts of possibilities latent within it, from the fire in the furnace to the shingles on the house, only one or the other of these can become actual. This is why Hegel emphasizes that conditions "go under" when the actuality "comes forth." If the fire were to become actual from the latent possibility in the wood, it would literally burn the wood to dust. While the shingles of the house still hold within them the real possibility of the fire, only one or the other of these can remain. If the fire takes the shingles in the night, this is because it has burned the wood out of them, and destroyed the form that they had become. There are of course various degrees of actualization. The shingles can become partly burned and remain partly intact, or in certain cases one condition can satisfy multiple possibilities at the point of actualization. But this only occurs when the contextual relationship happens to contain non-opposition and mutual exclusivity within its own content. And even this remains a unity of indifference and convenience.

Although they might seem to be the same at first glance, the problematic of *real disjunction* is significantly different from the principle of real contradiction that we had just outlined at premise nineteen. It is one thing to acknowledge the problematic of real disjunction, that each condition contains a multiplicity of real possibilities latent within it, but that only one or the other of these can become actual. It is quite another thing to acknowledge the conditionality underlying the conditions, which comes from real contradiction, that if something initial were to result in actuality, it would have to remain coherent throughout and not contradict the context that determines it. The apple (depending on its context) carries within it the possibility to be eaten, the possibility to be smashed, the possibility to hang on the tree; while it would be a contradiction if the apple were to be eaten and to remain hanging on the tree *in the same manner*, it would be a contradiction of a different sort (as per premise nineteen) if the apple were to become ingested by means that contradict the conditions of its ingestion.

(2) We can also demonstrate this contradiction in the very terms of how something initial results in actuality. I call this version the contradiction from developed actuality. On the one hand, what emerges is the same actuality throughout. The thing in question has become itself. The tree has grown from its possibilities in the seed. Someone's moral disposition has developed from the habits she had formed as a child. However, at the same time and in a contradictory way, the actuality that results from this is something that none of the moments alone can complete, and that not even the summation of all moments together can properly expose. The developed actuality is in this sense the actualization into an other, but this is equally the actualization into oneself. Again, this is why Hegel emphasizes that the conditions "go under" as the actuality "comes forth." Although they initially stand against the thing in question, as an external requirement that must be overcome, when the conditions are made present at the point of actualization, the possibility becomes realized and what began as only an external relation becomes a relation that is simultaneously external and internal.

This is not only a case of turning what is unlike into what is like. It is also a case of how to think of something as the other of itself. When the doe drinks the water of the forest, she digests the other. She makes it what she is. The water begins as an external contingent actuality. But because the actuality is contingent, it is not only what it seems to be. It is also the possibility of the doe. And the doe is not only what she seems to be. When she drinks the water, she turns this into what she is, but also only continues to be herself in this transformation. She sustains herself in the contraries of possibility. By taking up the possibilities of her conditions which exist dispersed in others, the doe becomes the other that she is and sustains herself in this.

This is why Hegel says that from the completion of the conditions, "the something itself is . . . determined as being equally actual and possible."[19] The actual that results from the actualization of conditions is not a new distinct actuality emerging against the background of many contrary possibilities. Nor is it an actuality that entirely succeeds the initial actuality of its conditions. What results is an actuality that is both itself and possibility. This means that the contradiction that Hegel had discovered at premise nineteen is not only restrictive, but also productive. The contradiction is restrictive in the sense that for the actualization to occur, one must follow the particular, determinate rules of one's conditions and not do otherwise. Yet the contradiction is at the same time productive in the sense that by satisfying the necessity of one's conditions, what comes of this is the realization of the contrary sides of possibility in one actuality. This is a reality that cannot be maintained formally. The real restriction that one must come to satisfy the conditions of one's possibility releases thought from that initial formal restraint, that something cannot both be itself and be the opposite of itself. Looking back at premise fourteen, we can now fully recognize the reason why real actuality *is* action—because real contradiction has put it in motion.[20]

21. Something initial becomes actual if all of its conditions are present. Therefore, what is really possible can no longer be otherwise. This possibility is real necessity.

> Die Negation der realen Möglichkeit ist somit ihre Identität mit sich; indem sie so in ihrem Aufheben der Gegenstoß dieses Aufhebens in sich selbst ist, ist sie die reale Nothwendigkeit. (WL 211)
>
> Was nothwendig ist, kann nicht anders seyn; aber wohl was überhaupt möglich ist; denn die Möglichkeit ist das Ansichseyn, das nur Gesetztseyn, und daher wesentlich Andersseyn ist. Die formelle Möglichkeit ist diese Identität als Uebergehen in schlechthin Anderes; die reale aber, weil sie das andere Moment, die Wirklichkeit, an ihr hat, ist schon selbst die Nothwendigkeit. Was daher real möglich ist, das kann nicht mehr anders seyn; unter diesen Bedingungen und Umständen kann nicht etwas anderes erfolgen. Reale Möglichkeit und die Nothwendigkeit sind daher nur scheinbar unterschieden; diese ist eine Identität, die nicht erst wird, sondern schon vorausgesetzt ist, und zu Grunde liegt. Die reale Nothwendigkeit ist daher inhaltsvolle Beziehung; denn der Inhalt ist jene ansichseyende Identität, die gegen die Formunterschiede gleichgültig ist. (WL 211)
>
> The negation of real possibility is therefore its identity with itself. Inasmuch as in its sublating it is thus within itself the recoiling of this sublating, it is real necessity.
>
> What is necessary can no long be otherwise. But in general what is possible can be otherwise. This is the case because possibility is an in-itself that is, at the same time, a positedness, and therefore essentially otherness. Formal possibility is this identity as transition into a sheer other; but real possibility, because it contains the other moment, actuality, is already itself necessity. Therefore, what is really possible is that which can no longer be otherwise. Under particular determinations and circumstances nothing else can unfold. Real possibility and necessity only seem to be distinct. They do not become this identity, but rather, their identity is already presupposed and lies in their ground. Real necessity is, therefore, a relation full of content, for the content is that identity, existing in itself, which is indifferent to form.

Real contradiction leads Hegel to claim that possibility and necessity only seem to be different. If by this he means that possible things are necessary, then this is one of the more provocative statements he makes in the chapter.

Although most of us would agree to the basic axiom that if something is necessary, then it is also possible, the claim becomes much more controversial if Hegel also mean that being possible makes something necessary. In the passage quoted above, Hegel acknowledges traditional definitions of necessity and possibility when he writes, "[W]hat is necessary can no longer be otherwise. But in general what is possible can be otherwise" (*Was notwendig ist, kann nicht anders sein; aber wohl was überhaupt möglich ist*). Why would he claim at this point in the argument that possibility and necessity only seem to be distinct?[21]

To analyze this question, I will explore three variations on the theme that if something is possible, this makes it necessary. Each variation clarifies how to think of the entailment from possibility to necessity without, however, falling into the logical form that most of us would find to be false, namely that what is merely possible in the limitless sense of formal possibility is necessary simply because it is possible. Each of the following variations presents an alternative that makes the entailment from necessity to possibility less problematic: (1) If something is possible, this makes it necessary, but not necessarily actual. (2) It is necessary that if something has become actual, it must have become actual through other actuals, that is, through conditions. (3) It is necessary for a possibility to become actual, but not necessarily in any specific way.

(1) If something is possible, this makes it necessary, but not necessarily actual. This variation assumes that all possibilities of a given determinate range necessarily exist, and that if any one strand of this range were not to exist, the actualization could not occur. Now, if this is right, then we will need to posit a distinction, which Hegel did not give a full account of in the formal argument, between existence and actuality. Although only some possibilities become actual, all possibilities exist. This is then a development upon the insights of premise ten.

Hegel's language of "presence" and "completion" suggests that all determinate possibilities must exist *prior* to their actualization and that this actualization depends upon the more original existence of unactualized possibilities. This variation on the claim works well in the sense that we are able to establish the entailment from possibility to necessity while avoiding the controversy that mere possibility necessarily entails actuality. However, I believe that this variation will fit better in the absolute modality steps of Hegel's argument, since there we discover an actuality that retains the existence of unactualized possibility as part of its constitution. At this stage, the entailment from possibility to necessity remains exterior to the actualization, which is construed more narrowly because this requires us to posit a distinction between existence and actuality, a distinction that will fall away in absolute modality.[22]

(2) It is necessary that if something has become actual, it must have become actual through other actuals, that is, through conditions. This variation exposes the necessity in the movement from conditions to further actualities. Hegel claims that if something is a condition for something else, then it is an immediate actuality whose content as actual entails the possibility of other actuals. In this variation, we define real necessity as the conditionality of the condition. If something initial results in actuality, this is because all of its conditions have become completely present.

At premise eighteen, thought discovers material actualization, where something initial comes into actuality only if all of its possibilities (as conditions) are made present. The advancement at premise twenty is to realize the conditionality behind this development from the initial to the resulting actuality. Something initial can only result in actuality if it can complete itself in others. This is no longer merely a claim about possibility, where something finds itself dispersed in the possibilities of others. This is now a claim about necessity.

An engine is not completely an engine if all of the conditions that make it work are not already in place. The cylinder, the crankshaft, the sump, the spark plugs—these are ancillary conditions. They are possibilities that can no longer be otherwise because they are integral to making the engine run. If one of these parts were to go missing, we might say that the engine is broken or does not work properly, but we would also mean that the engine is not completely itself unless it contains the full range of its possibilities, both in the sense of the conditions that make it possible, but also in the sense that it must be able to perform the full range of its capacities (that is, burn fuel, produce motion, and so on). Real necessity is in this sense compulsive necessity. Once actualization has been initiated, something can no longer do anything at all but follows the course of its possibilities since these have become necessary.

This variation can offer us only an indirect version of the entailment from possibility to necessity, where possibility is necessary but only because it is immediately and contingently an actuality. For this variation to work, we must rely on a more basic connection between actuality and necessity, one where conditions are immediate actualities whose possibility entails other actuals. This means that the further entailment from possibility to necessity presupposes that real actualities necessarily lead to other actuals, that is, to conditions and results.

(3) It is necessary for a possibility to become actual, but not necessarily in any specific way. This variation explains why something possible is necessary by proposing that although possibility must become actualized, there is no one determinate possibility that must become actual any more than any other. The moment of necessity in this variation appeals to the basic assumption that there must be something rather than nothing. But the variation also

avoids the controversy that what is merely possible is necessary, or similarly, that only certain determinate possibilities must become actual. Its strength relies on a level-distinction between two types of determinateness, one where something must be determinate because being is, but also where there is no specific determinateness that should or must become actual. Since actuality immediately appears before us as the fact of existence (as per premise one), actualization cannot not happen; but this necessity where actualizing possibility cannot not happen makes no further claim to determinism.

Or, Hegel might mean the more contentious claim that to posit a given determinate range of possibilities commits something of this range to become actual. This combines variation (2) and variation (3) in the sense that although there is no specific direction the condition must take, that it is a condition at all commits it to some further result in actuality, even if this is the actuality of failed conditions or of sheer, accidental contingency. A claim like this one is more contentious because of its proximity to the claim that if something is merely possible, then it must become just that determinate actuality and nothing else (as if the mere possibility of the unicorn were to necessitate actual unicorns). But if Hegel is saying that only something of a given range of possibilities must become actual, but not that any specific possibilities must come about, then the distinction still holds.

Hegel probably has all three of these variations in mind when he says that what is really possible can no longer be otherwise. We can read variation (1) as the most conclusive of the three variations since it relies on the theory that unactualized possibilities exist and are constitutive of the contextual world. However, this variation also reaches too far ahead of Hegel's argument into the terrain of absolute modality, drawing us prematurely to the theory that unactualized possibilities not only exist, but are a necessary component for the absolute actuality of substance. I think that variation (2) is the most suitable candidate for the state of affairs of premise twenty-one, since it shows the necessity embedded in the possibilities of conditions. Here, possibility is only indirectly necessity, but this is probably right, as Hegel suggests by his choice of words "possibility and necessity are only seemingly distinct" (*Möglichkeit und die Notwendigkeit sind daher nur scheinbar unterschieden*). Variation (3) should seem more appropriate to the following premise, where Hegel claims that contingency stands at the base of real necessity.

22. Contingency is nevertheless the reason behind why what is really possible can no longer be otherwise.[23]

> Diese Nothwendigkeit aber ist zugleich relativ.—Sie hat nemlich eine Voraussetzung, von der sie anfängt, sie hat an dem zufälligen ihren Ausgangspunkt. Das reale Wirkliche als solches, ist nemlich das bestimmte Wirkliche, und hat

> zunächst seine Bestimmtheit als unmittelbares Seyn darin, daß es eine Mannichfaltigkeit existirender Umstände ist; aber diß unmittelbare Seyn als Bestimmtheit, ist es auch das Negative seiner, ist Ansichseyn oder Möglichkeit; so ist es reale Möglichkeit. Als diese Einheit der beyden Momente ist sie die Totalität der Form, aber die sich noch äusserliche Totalität; sie ist so Einheit der Möglichkeit und Wirklichkeit, daß 1) die mannichfaltige Existenz unmittelbar oder positiv die Möglichkeit ist; - ein mögliches, mit sich identisches überhaupt, darum weil sie ein wirkliches ist; 2) insofern diese Möglichkeit der Exstenz gestzt ist, ist ie besimmt als nur Möglichkeit, als unmittelbares Umschlagen der Wirklichkeit in ihr Gegentheil, - oder als Zufälligkeit. (WL 211)

> However, because it has a presupposition from which it begins, that is, because it has its starting point in contingency, this necessity is at the same time relative. As such, real actuality is determinate actuality, and at first its determinateness, as an immediate being, is in terms of a multiplicity of existing circumstances. But this immediate being as determinateness is also its own negative, the in-itself, in other words, possibility. In this way, it is real possibility. Because it is the unity of these two moments, it is the totality of form, but a totality which is still external to itself. It is a unity of possibility and actuality in the following ways: (1) multiplicity existence is possibility immediately or positively—because it is actual, it is possibility and is something self-identical as such. (2) Inasmuch as this possibility of existence is posited, it is determined as only possibility, as the immediate conversion of actuality into its opposite—or as contingency.

If something is really possible, this makes it necessary. But directly after saying this, Hegel claims that the necessity in this is at the same time relative because it has its starting point in contingency. Now, he probably has a traditional definition of hypothetical necessity in mind, such as Aristotle presents in the *Physics* 2.9 when he says that "in all . . . things which involve that for the sake of which: the product cannot come to be without things which have a necessary nature, but it is not due to these (except as its material); it comes to be for an end" (CWA 200a7–10). In terms of conditional propositions, real necessity appears in the relationship between the antecedent and the consequent, but this relationship nevertheless has contingency at its base because one does not need to take up the conditional in the first place. In other words, even if we assume that B must follow necessarily from A, it is still contingent whether one presupposes A or not. The engine burns gas and produces motion, but only if the conditions that make it function are already in place. Let us assume they are in place. Then the engine cannot not function. However, to assume this is to expose the relativity of real necessity. Because the conditions could have been otherwise, what follows of necessity is at the same time relative and therefore contingent. Relative necessity is, in effect, goal-oriented necessity.

One issue that arises from this reading of relative necessity is that it alludes to immediate freedom from determinateness. All things would seem to have the option not to be determinate, some in the active sense that they might not choose to become what they already tend towards being (for example, a talented child might not choose to become a professional pianist), others in the passive sense that it would be a mere contingency of chance whether they come about or not. But I doubt Hegel means freedom from determinateness in this respect. He seems to emphasize, to the contrary, that from the disposition of the determinateness, the thing cannot but take up the conditions that determine it. Rather than the sign of freedom from determinateness, Hegel probably intends real contingency to be the sign of instability. Because we must assume the conditionality of the conditions, there is the risk that something might not become what it already tends to be.

The argument for this requires us to rethink the stability and identity-with-self of what is immediately actual. It is not simply the case that from material actualization, something can or can not come to actualize the contraries of possibility in one actuality. Hegel's point is that if something does not maintain the contraries of possibility, then it cannot maintain itself. Its own identity fundamentally depends upon the actualization of possibilities that are initially in others. He claims that real necessity is relative, but not because something has the option to become itself in others, and the equal option not to do this. His point is the opposite. If contingency stands at the base of real necessity, then material actualization is the risk that something might not become itself because it cannot meet the requirements for the possibility of its activation in and as others. The relativity of real necessity is not freedom from determinateness but the requirement that something must sustain the contraries of possibility to become itself.

One reason why we should not read freedom from determinateness into Hegel's claim is because then we would have to presuppose the determinateness and stable identity of things from a position prior to their actualization. Although the prior stable identity might seem intuitive, the question of whether something initial will become actual is a question that can only appear after the actualization has already occurred. Contingency is then the starting point of real necessity because none of the conditions alone can anticipate what the thing in question is that will result from them. Actualization requires retroactive presupposition. If what is immediately actual is identified in its possibility to-be, its identity only becomes stable in the instability that it equally can not be. This is just the superficial confirmation of something's self-coherence through the law of identity, that if A is possible, then A is A. But there is also a significant implication from this. If the doe must confirm the actuality of being a doe in the possibility of not being a doe, through the conditionality of whether her conditions will be met, she is not always already what she is but only becomes herself from the contingency that she must also be able not to be this. That she is what she is presupposes that she can also not be what she is. Determinateness necessarily requires the negation of the determinateness as the more originary position from which something determinate can be. Real necessity does not present things as predetermined to be what they are, as the connotations of hypothetical necessity sometimes suggest. Rather, it shows us that the stable identity of things lies in the more originary relation of contingency. Something must have been able not to be what it is in order to be what it is.

NOTES

1. Stekeler-Weithofer says that to do justice to the connotations of action and work associated with Hegel's description, it would be better to translate *energeia* and *Wirklichkeit* as "reality" rather than actuality (MN 238). Also see Pirmin Stekeler-Weithofer, "Scientific Truth as Augmented Reality: On the Contrast between 'Wirklichkeit' and 'Actuality,'" *De Gruyter* (2017): 83–100.
2. For Hegel's account in the *Logic* of "the thing with many properties," see WL 327–34, SL 484–92. Cf. Hegel's discussion in the *Phenomenology*, 67–79 (paragraphs 111–31).
3. See "(a) Something and an Other" in WL 67–69, SL 117–22.
4. Hoffmeyer says this quite well. The ought is both the prevention of the other but also an elevation into the other. Hoffmeyer, *The Advent of Freedom*, 30–31.
5. Ferdinand de Saussure, *Course in General Linguistics* (New York: McGraw-Hill Paperbacks, 1966), 144.
6. De Saussure, *Course in General Linguistics*, 116.
7. In his analysis of real possibility, Yeomans explores a very helpful detail in what he calls "modal metrics." He defines modal metrics as a "standard for measuring the possibility of an event." The more thought comes to recognize the real conditions of something, the more accurate thought can become at exposing the real possibilities of something. Yeomans sees in Hegel's transition from formal to real modality the insight that we can look more or less closely at the background circumstances of an event, and establish from this a standard for measuring what is possible and what is not possible *relative to the context*. Yeomans gives the example of how to measure whether it is possible to run a four-minute mile: "It is logically possible for me to run a four-minute mile. If we next consider not all the relevant circumstances, but only some very general ones, it might still be possible for me to run a four-minute mile. So if we only consider the fact that I am a human being with lungs and legs and eyes to guide myself, and the fact that some human beings can run a mile in under four minutes, then it is perhaps still possible that I could run a four-minute mile. But if we then consider more background conditions like my lack of training and rickety knees, it becomes impossible that I could run a four-minute mile. . . . We can make modal judgments at any point along a continuum that runs from considering no background conditions (logical modality) to considering all background conditions (absolute modality), but the significance of our judgments varies directly with the extent of background conditions considered." Yeomans, *Freedom and Reflection*, 151.
8. I realize that by invoking this analogy between Heidegger's contextual world and Hegel's existing multiplicity, I present the *Logic* as more phenomenological and experiential than it might appear to be. The limitless multiplicity of the formal passages might indeed appear as a primitive logic that is essentially different from the experiential outlook of Heidegger's temple. But I maintain that Hegel's analysis of real modality requires phenomenological examples about contextual experience.
9. Martin Heidegger, "The Origin of the Work of Art" in *Poetry, Language, Thought*, translated by Albert Hofstadter (New York: Harper & Row, 1971), 42–43.
10. Heidegger uses this phrase in a related example about the Aegina sculptures being transplanted to the Munich art collection. Heidegger, *Poetry, Language, Thought*, 40.

11. Casey's extensive conceptual analysis of place supports this analysis. See Edward S. Casey, *The Fate of Place: A Philosophical History* (Berkeley: University of California Press, 1997) and Edward S. Casey, *Getting Back into Place: Toward a Renewed Understanding of the Place-World* (Bloomington: Indiana University Press, 1993). In *The Fate of Place*, Casey says that one of the most important questions Heidegger asks in *The Origin of the Work of Art* is not about "what the work of art *is*" but rather "where it belongs." "The what-is question," Casey writes, "leads to a false essentialism, to mere definitions and formal features. The question as *to where* leads us straight to the work of art itself: to where it exists as a scene of primal conflict and unconcealment." Casey, *The Fate of Place*, 265. Casey emphasizes that the Greek Temple is the place where truth happens. However, in *Getting Back into Place*, Casey argues that the *lived body* is conspicuously missing from Heidegger's account of the Greek Temple. If we were to also focus on the lived body, we would then recognize that there exist place-modal relations, "relations of inside-outside, alongside-around, and with-between," and not only the site of the building and its environment. Casey, *Getting Back into Place,* 131–32. Lived bodies are a necessary condition for what Casey calls "leeway," not to be measured by space-terms such as "feet, inches, or *braccia*," but by place-terms as an existing multiplicity of relationality. Hegel also does not venture into the topic of lived bodies and the kinds of modal relations that might come from this, a topic which might seem conspicuously absent from these passages of the *Logic*. For an argument that the lived body is an important theme in the *Phenomenology of Spirit*, see John Russon, *The Self and Its Body in Hegel's Phenomenology of Spirit* (Toronto: University of Toronto Press, 2001).

12. See "The Relation of Causality" of "The Absolute Relation" chapter (BK 2, SN 3, CH 3, DV 2), which follows directly after the "Actuality" chapter, for Hegel's discussion of the cause-effect version of dispersed actuality. Here, Hegel discusses the cause-effect version of "dispersed actuality." He writes, "[T]he actuality which substance has as cause, it has only in its effect" (WL 397/ SL 559). The cause only becomes actual insofar as it finds its resolution in the effect. This means that the cause itself, although it is truly the actuality, depends upon the effect because it is itself dispersed in it.

13. Hoffmeyer notices that this sentence has an earlier iteration in the "Existence" passages of the *Logic*. Hoffmeyer, *The Advent of Freedom*, 31. "When all the conditions of a fact are present, it enters into existence" (SL 477).

14. Ng, "Hegel's Logic of Actuality," 11.

15. In the *Lectures on Logic*, Hegel calls these the three moments of necessity. G. W. F. Hegel, *Lectures on Logic,* translated by Clark Butler (Bloomington: Indiana University Press, 2008), 161–64.

16. In his inquiry about the relationship between equipment and art, Heidegger also discusses the distinction between "using" and "using up." He says, for example, that when we make an axe from stone, we use the stone up in the sense that it disappears into what the axe is. However, in his famous example of the Greek Temple, Heidegger claims that the material of the temple is not "used up" in the same way, because the material lets the world of the temple "come forth," and in this sense, the material does not "disappear." Heidegger, *Poetry, Language, Thought*, 44–45.

17. Longuenesse's analysis of the role of activity is generally excellent. She divides activity into both agency (for example, "a man, a character") but also the movement from conditions to the thing (CM 151–52). She claims that activity is the main transition-point from real necessity to absolute necessity, and from this into the concept (*Begriff*). In her conclusion, she supplies us with reasons for why Hegel's concept of activity is a direct development of the Kantian "I think" (CM 160–62).

18. Longuenesse makes this same point when she warns against inserting temporality into the process between conditions and the resulting actuality (CM 137–38).

19. WL 387, SL 548. For my discussion of this sentence, see premise eighteen.

20. When I claim that to become actual is to become in contradiction, or that this version of contradiction is a contradiction *of movement,* I have Hegel's remarkable statement from the contradiction passages of the *Logic* in mind: "As against contradiction, identity is merely the determination of the simple immediate, of dead being; but contradiction is the root of all

movement and vitality; it is only in so far as something has a contradiction within it that it moves, has an urge and activity" (WL 286, SL 439).

21. Commentators have addressed this controversy in the chapter, but none have explicitly situated it in terms of real contradiction. Di Giovanni sees the necessity of real possibility as the inevitable and irreversible passage from possible conditions to the resulting actuality (CC 190). Marcuse calls this unity of possibility and necessity a type of movement. "*Necessity* makes up the character of this movement. For actuality only unites itself with itself through its movement, and that into which it is "transformed" is no other than its own possibility, its own proper being (*Ansichsein*)." His emphasis on the self-movement aspect of Hegel's claim is quite helpful, but he does not address whether real contradiction is the source of this movement (HO 96). Carlton says something similar about necessity and movement when he defines real necessity as an event, rather than a thing, principle, or restriction. This is helpful for thinking of real necessity as an element in the process of actualization. Carlton, *A Commentary on Hegel*, 409. Lampert makes a larger claim to this effect when he says that "necessity" is generally about multiplicity for Hegel: "Readers sometimes err in thinking of necessity in Hegel as a kind of identity rather than a kind of multiplicity." The three variations of possibility entailing necessity that I will present in this premise are each already anticipated in Lampert's essay (FM 75).

22. Although he does not directly say that all determinate possibilities of a range must exist, we can still take Houlgate's point about war and peace as an example of variation (1), in the sense that if peace is to become actual, the possibility of war must also exist. Houlgate focuses on the claim that what is really possible must be replaceable with another, and that in this way its contingency exists. However, I maintain that the difficulty at this stage of Hegel's argument is to recognize why possibility entails necessity. Cf. Houlgate: "What becomes clear in section B is that, for Hegel, the actual possibility of being *or* not being always takes the form of the *real* possibility of something else, that the sheer contingency of something is expressed in the fact that it itself contains the possibility of its being replaced by another. Peace is merely contingent, therefore, to the extent that it itself contains the possibility of war" (NC 43).

23. It is worth noting that in the formal sub-chapter, Hegel discovers formal possibility from formal actuality, then contingency from formal possibility, and then necessity from contingency; however, in the real sub-chapter, while he continues to deduce real possibility from actuality, Hegel now claims that necessity comes from possibility, and contingency from necessity. In other words, in the real modality sub-chapter, Hegel reverses the order of the deduction between contingency and necessity.

Chapter Three

Absolute Modality

23. Absolute actuality is an actuality of the entire process once all of the conditions are present. It is the unity of contingency and necessity.

> Daher ist diese Möglichkeit, welche die unmittelbare Wirklichkeit, indem sie Bedingung ist, an ihr hat, nur das Ansichseyn als die Möglichkeit eines Andern. Dadurch daß, wie gezeigt, diß Andersseyn sich aufhebt, und diß Gesetztseyn selbst gesetzt wird, wird die reale Möglichkeit zwar Nothwendigkeit; aber diese fängt somit von jener noch nicht in sich reflectirten Einheit des Möglichen und Wirklichen an;—dieses Voraussetzen und die in sich zurükkehrende Bewegung ist noch getrennt;—oder die Nothwendigkeit hat sich noch nicht aus sich selbst zur Zufälligkeit bestimmt. (WL 212)
>
> An sich ist also hier die Einheit der Nothwendigkeit und Zufälligkeit vorhanden; diese Einheit ist die absolute Wirklichkeit zu nennen. (WL 213)
>
> Therefore, this possibility, which immediate actuality has as its condition, is merely the in-itself as the possibility of another. Because of this, as we have indicated, the in-itself sublates itself, positedness is thereby posited, and real possibility becomes necessity. But this necessity begins from a unity of possibility and actuality that is not yet self-reflected. These presupposing and self-returning movements are still separate. In other words, necessity has not yet of its own accord determined itself as contingency.
>
> The unity of necessity and contingency is present here in principle. This unity is called absolute actuality.

Contingency is the foundation for real necessity. Contingency's role at this point in the argument exposes the instability of actuality and causes Hegel to turn to absolute actuality. Real actuality is the manifestation of this or that determinate content. But with absolute actuality, Hegel has made explicit what he has already begun to articulate, the large actuality that is itself the total form of possibility, the entire state of affairs surrounding the content of each determinate thing.

If his argument were to end at premise twenty-one, it would seem as if real necessity were to come from an external source. Although conditions would remain the unity of the actual and the possible, we would only be able to comprehend the reason or ground for these conditions from a standpoint beyond the process of the conditions. Hegel first attempts to establish the reasons for material actualization from beyond the process of the actualization. Why does the plant grow from the seed? Because the soil is rich, the water is plentiful, and the sunlight is strong.[1] This attempt at exteriorizing the reasons for actualization is no doubt a residual effect from the initial formal problem, that actualization cannot actualize the contrariety of possibility without falling into contradiction or disjunction. But such external explanations lead only to a kind of infinite regress where each condition supplies the reason for another, *ad infinitum*. Ultimately, if the reason for the manifestation comes from a source exterior to the process of the manifestation, then there is really no good reason behind the necessity of the conditions.

Hegel then proposes instead that although the reason behind the resulting actuality might seem to come from a source external to it, this source is really immanent to the resulting actuality. Actuality must supply its own immanent reasons for its manifestation. If the reason for the manifestation can only be found in another, the rationality behind the actualization remains exterior to the actualization and contingency remains separate from it. However, if the reason for the manifestation is immanent to the conditions and to everything that develops from them, then the necessity has supplied its own sufficient reason. Then the actual is the cause of itself. Its movement in possibility is the movement of itself. But this requires a significantly different type of modality. We must now think the identity of actuality as a circuit or a process rather than as a thing in the context of others. This process-actuality is the absolute actuality of self-movement.

Absolute actuality is the actuality that emerges from the conditions of possibility when all of the conditions are completely present, not only in the hypothetical sense of relative necessity beginning from contingent external actualities, but in the sense of total inclusiveness, where necessity and contingency are in principle the same. Real actuality took its possibilities to be equally determinate, external requirements, existing initially in other actuals, but one recognizes in absolute actuality that the possibilities are the self-movement of the same actuality throughout. In this sense, absolute actuality is the realization of real actuality. Real actuality only entertained the alien character of possibility so that this actuality could become itself. When real actuality overcame the conditions of its possibility, it included the other within itself. The conditions released themselves of externality, and what was initially possible became itself in the other. Absolute actuality is then the realization that real possibilities only seemed to be dispersed in other actuals, but that this process of satisfying external conditions is the self-same movement of actuality throughout. This only works if the actuality includes the other as the constitution of itself (and if the other includes the self as the constitution of the other).

Hegel's theory of absolute actuality is in this sense closely related to Leibniz's theory of compossibility. In paragraph sixty-one of "The Monadology," Leibniz describes the relation of composites. "Every body," he writes "is affected by everything that happens in the universe, to such an extent that he who sees all can read in each thing what happens everywhere, and even what has happened or what will happen."[2]

While real modality invokes images of production, ground, and the movement between self and other, now Hegel invokes images of a different sort: tracings (*Spur*), markings (*Maal*), reflexes (*Reflex*), absolute negativity (*absolute Negativität*), and freedom (*Freyheit*). In premises twenty-four through twenty-seven, let us examine the consequences of this third type of modality, a modality where actuality can act but only upon itself, and where necessity and contingency turn out to be the same.

24. Necessity, not possibility, is the reflection of absolute actuality into itself.

> Diese Wirklichkeit, die selbst als solche nothwendig ist, indem sie nemlich die Nothwendigkeit als ihr Ansichseyn enthält, ist absolute Wirklichkeit;—Wirklichkeit, die nicht mehr anders seyn kann, denn ihr Ansichseyn ist nicht die Möglichkeit, sondern die Nothwendigkeit selbst. (WL 213)

> This actuality, which is itself as such necessary, since it contains necessity as its in-itself, is absolute actuality—an actuality which can no longer be otherwise, for its in-itself is not possibility but necessity itself.

Throughout the argument, Hegel has consistently called the in-itself (*Ansichseyn*) possibility, not necessity. At premise three, he called possibility the in-itself of actuality because it reflects actuality into itself. Likewise, real possibility is the in-itself of real actuality because an existing thing can only manifest itself by realizing its possibilities in others. This possibility is, as Hegel says, full with the content (*Inhaltsvolle*) of other actuals. However, the structure of absolute actuality works quite differently. Whereas formal and real actuality each in turn contain formal and real possibility as the in-itself, absolute actuality contains necessity straight away as the in-itself.

Notice what an unusual view of necessity Hegel maintains at this stage of the argument. While he does not yet call this absolute necessity, we can already anticipate the reason why this actuality can no longer be otherwise. Since there is no longer the possibility of an other that the actuality is not, there is no longer any way for the actuality to be other than it is. Therefore, this actuality is the manifestation of necessity.

And yet the reason why this actuality cannot be otherwise is because it is already the other of itself, and this other is already what it is. Since absolute actuality constitutes the disposition of the entire set of conditions, it embodies all permutations of possibility. This is an unusual usage of necessity because it comes from inclusion rather than exclusion. Actuality includes what had been for formal and real actuality the non-actual.

At premise twenty-three, we described absolute actuality as a process or a circuit. If the actual embodies all permutations of possibility, it reveals the identity of the relations between things in their context. It is, for example, the process between the seed, its conditions, and the plant that it becomes. The actuality resides in every frame of its growth and in every condition that it requires; and yet although it resides in one frame, this does not exclude its being at the same time the actuality of all the other frames, the totality as one disposition. But by emphasizing the necessity of this actuality, Hegel exposes something more at premise twenty-four than process, circuit, or relation. That necessity is the in-itself reveals the universality of absolute actuality. Universality, in this modal sense, means that the negativity of the possible is expressed completely in the actual. With every instantiation, the actuality remains itself because it is the universality of each possibility as one set. To think absolute actuality as the total inclusion of all possibilities in a set is to think the large actuality of genus,[3] but with one important qualification. Just as this actuality is the total inclusion of each individual within the set, it is at the same time singular determinateness.

In this sense, Hegel probably has Aristotle's distinction between first- and second-order actuality in mind. In *de Anima* 2.1, Aristotle says that there are two kinds of actuality, one primary and one secondary (CWA 412a1–413a10). Second order actuality presents what an agent is actually doing at the point when it is doing it. For example, at this moment, I am writing at my desk. This activity is derivative because it is not as if I lose the ability to write at my desk whenever I stand up and cease to write. Aristotle calls secondary actuality the "waking" side because it is the present position of ones activities. Although in some statements Aristotle calls primary actuality the "sleeping" side, he really means that primary actuality is both sleeping and waking, since this type of actuality covers the whole of an organism's abilities and being. For Aristotle, substances are primary actualities because they simultaneously sustain all contraries without turning contraries into contradictories, but equally, without losing the coherence of their identity. If we were only able to give an account of living substances in terms of secondary actuality, we would not have the resources to explain how substance sustains contrary dispositions without becoming contradictory.

The horse galloping in the field is an example of absolute actuality, but only if we recognize that essence and existence are the same. Horseness embodies all of the possibilities of being a horse. She has no way to exceed herself in this. She is completely consumed in being a horse. Absolute actuality demonstrates that there is no possible way for her to be otherwise than to be what she is. Each instantiation of what she is can be nothing other than her essence. This is because every possibility of being a horse is necessary to horseness. This does not mean that the set is finite or exhaustible. There is a certain inexhaustible variety to her genus, which no one action could completely anticipate. And yet, at the same time, the horse is not merely a representation or instantiation of horses; she appears as her essence, not as a substratum standing below or beyond what she seems to be, but as the whole possibility of what she is. This is why Hegel says that necessity is the in-itself of absolute actuality: because any possible permutation was already necessary to the set. Since absolute actuality is the actuality of itself and its other, any further permutation of the other is still the self-same actuality (and any further permutation of the self is, likewise, the other that is the self).

Absolute actuality still faces a significant problem, one that leads Hegel from absolute actuality to absolute possibility, and leads us to the further distinction from necessity as the in-itself to absolute necessity. The problem with thinking genus-actuality is that, while it includes all possibilities of its kind, it cannot include contingencies. These, instead, fall outside of the set of absolute actuality. The contingency of death turns the living horse into a corpse. Likewise, while a horse with only one leg might remain within the genus of horse, if the horse were to become reorganized in an even more radical fashion, it would eventually reach the threshold of contingency, and fall outside of its genus altogether into formless matter. While defining absolute actuality as genus-actuality is a good way to begin to think the logic of absolute modality, if Hegel really means what he says—that absolute actuality literally embodies every permutation of possibility—he will need to explore an even larger version of actuality, one that can include all contingences that would seem to fall beyond it. But to include contingency is to approach a difficult thought. This is the thought of total inclusion from which there is no remainder, the absolute conversion of actuality and possibility.

25. Necessity is the in-itself because absolute actuality already includes all possibilities. This actuality is as much possibility as it is actuality. Hegel calls this possibility absolute possibility.

> Aber damit ist diese Wirklichkeit, weil sie gesetzt ist, absolut, das heißt, selbst die Einheit ihrer und der Möglichkeit zu seyn, nur eine leere Bestimmung;

> oder sie ist Zufälligkeit.—Diß Leere ihrer Bestimmung macht sie zu einer blossen Möglichkeit, zu einem, das eben so sehr auch anders seyn und als Mögliches bestimmt werden kann. Diese Möglichkeit aber ist selbst die absolute; denn sie ist eben die Möglichkeit, eben so sehr als Möglichkeit wie als Wirklichkeit bestimmt zu werden. Damit, daß sie diese Gleichgültigkeit gegen sich selbst ist, ist sie gesetzt als leere, zufällige Bestimmung. (WL 213)

> However, because this actuality is posited as absolute, that is, as the unity of itself and possibility, it is only empty determination. In other words, it is contingency. This emptiness of its determination makes it into a mere possibility, one which can just as likely be otherwise than it is and can be determined in terms of possibility. But this possibility is itself the absolute, since it is a version of possibility that can be determined equally by possibility or actuality. Because it is indifferent to itself, it is posited as an empty, contingent determination.

The reader might conclude from the description Hegel gives of absolute actuality and from the connotations associated with the adjective "absolute" that there is no good place for the other, or for difference, in Hegel's modal theory. But his advancement from absolute actuality to absolute possibility shows that this is not the case. Hegel's point is not simply that in absolute actuality the other has become completely subsumed within the one comprehensible actuality, in effect dissolving the other of its true otherness. We can find this sort of objection to Hegel in the first chapter of Levinas's *Totality and Infinity*. Levinas defines ontology as an approach to knowledge that attempts to uncover what the other is in fact and in its totality, so that one might claim to know what the other is and what the other can be. Levinas rejects this approach because he sees in it an inappropriate seizure of the Other, a reduction to factual knowing of precisely that which cannot be known. In contrast to this, Levinas defines metaphysics as a non-knowledge based approach that attempts to let the other be the appearance of an infinite, inexhaustible, and ungraspable beyond.[4] Although Hegel's absolute actuality might seem to follow the ontological approach as Levinas outlines it, I argue that because of absolute possibility, Hegel's intention is to posit an actuality that lets the possible *qua* the possible appear, without, however, exhausting the possible or rendering it finite.[5]

Absolute actuality yields a concept of otherness that has become able to go free from the limitations of its determinateness. Hegel discovers the concept of possibility *qua* possibility, an other that is the negation of the limitations that had produced it as an other. Absolute possibility is a negation of real determinateness, which is already a negation of the form. In this sense, absolute possibility is the negation of the negation.

Absolute possibility significantly expands absolute actuality. Although a given genus contains every possible permutation, even an infinite variety of permutations, from within its membership, a genus must still follow the rules of exclusive necessity in the sense that if, for example, the horse is a horse, this necessarily excludes in a restrictive sense the possibility of becoming-cow or of becoming-umbrella, in other words, the possibility of becoming something that exists outside of the exclusive membership. However, because of absolute possibility, each individually existing thing has become a perspective of everything, as with the monad, an actual that is at the same time the whole possibility. Once Hegel advances to absolute possibility, he requires the necessity of total inclusion, the necessity of the possible itself. The transition from premise twenty-four to premise twenty-five is essentially the transition from a theory of many substances (such as Aristotle's) to a theory of one substance (such as Spinoza's). Paradoxically, possibility *qua* possibility becomes viable in Hegel's theory only when the possible becomes completely included in the constitution of the actual. This is why Hegel turns to the absolute necessity of being.

26. If absolute actuality is absolute possibility, the necessity of this actuality becomes absolute necessity.

> Die absolute Nothwendigkeit ist also die Wahrheit, in welche Wirklichkeit und Möglichkeit überhaupt, so wie die formelle und reale Nothwendigkeit zurükgeht.—Sie ist, wie sich ergeben hat, das Seyn, das in seiner Negation, im Wesen, sich auf sich bezieht und Seyn ist. Sie ist eben so sehr einfache Unmittelbarkeit oder reines Seyn, als einfache Reflexion-in-sich, oder reines Wesen; sie ist diß, daß diß beydes ein und dasselbe ist.—Das schlechthin Nothwendige ist nur, weil es ist; es hat sonst keine Bedingung, noch Grund.—Es ist aber eben so reines Wesen, sein Seyn ist die einfache Reflexion-in-sich; es ist, weil es ist. Als Reflexion hat es Grund und Bedingung, aber es hat nur sich zum Grunde und Bedingung. Es ist Ansichseyn, aber sein Ansichseyn ist seine Unmittelbarkeit, seine Möglichkeit ist seine Wirklichkeit.—Es ist also, weil es ist; als das Zusammengehen des Seyns mit sich, ist es Wesen; aber weil diß Einfache eben so die unmittelbare Einfachheit ist, ist es Seyn. (WL 215)

> Absolute necessity is the truth in which actuality and possibility generally, as with formal and real necessity, withdraw. It is, as we have seen, being, but being which in its negation, in its essence, nevertheless references itself as being. It is equally simple immediateness, in other words, empty being, as simple self-reflection, essence. It is both of these together. Absolute necessity is only because it is; it is neither condition nor ground. But, equally, it is pure essence; its being is simple self-reflection; it is because it is. As reflection, it has ground and condition, but it has only itself as ground and condition. It is the in-itself, but its in-itself is its immediacy; its possibility is its actuality. It is, *therefore*, because it is. As the going-together of being with itself, it is essence. But because this simple is immediate simplicity, it is being.

Hegel returns to the claim he had made at premise twenty-four, that necessity is the in-itself of absolute actuality, but now he calls this necessity absolute necessity. Once thought discovers that the possibility of absolute actuality is absolute possibility, this actuality can be nothing less than the actuality of the possible itself.

When Hegel discovers absolute necessity, he restates the concept of being. Being is the "whatever" of the is. It can include anything and everything within it. In this sense, the actualization of being contains possibility in the most extensive sense. Absolute necessity is being because being is that which nothing could exceed. There is no way around being, no possible remainder or outside.[6] In this sense, being is absolutely necessary and could not be otherwise. However, it would be misleading to say that Hegel has simply returned to the concept of being from the very beginning of the *Logic*, since this would bring along with it the connotations of indeterminateness, emptiness, and stasis, which I doubt Hegel intends. Although he claims that everything is formally possible, when he says that absolute necessity is both pure being and pure essence, he most likely already means the one substance, and means this in the most inclusive of ways. The consequence of absolute necessity, as we will see, is that this disposition must include in actuality the existence of unactualized possibilities as part of the constitution of what substance is, in the sense that whatever something becomes, it still is. This consequence is also becoming as self-movement, since Hegel also seems to be suggesting that, if something is a disposition of everything, it must become otherwise in order to be itself. This means that unactualized possibilities cannot simply retain their initial status as the unactual, but must also expand what it means to be actual, and come to include themselves in the actual, however in a different sense from the immediate surface actuality at premise one.

Hegel's argument that absolute necessity is a disposition of everything comes from his theory that being relates only to itself. He says this when he writes, "[Absolute necessity] is, as we have seen, being, but being which in its negation, in its essence, nevertheless references itself as being" (WL 215). If being were not related only to itself, the reason why being *is* would stand outside of being. Then there would be an external source for being. Then we would say that being is because of another, because of God or the sovereign or nature. But this would be absurd. If the reason for being were something other than being itself, the reason for being could not be. If we were to claim that God is the reason for being, then either we would need to admit that God *is*, in which case this simply affirms that being is self-relation, or we would need to admit that God is something other than being. But if the latter were true, then God would have no authority as the reason for why being is since God would not exist.[7]

Absolute necessity is both the most restrictive and the least restrictive type of necessity. It is the most restrictive because it compounds the restrictions of both formal and real necessity. Formally, anything is possible that does not contradict itself. Even under the extra restrictions of what is really possible, there still remain real alternatives and various ways to actualize the possible, an openness to anything from within the restrictions of content. But when it comes to absolute necessity, Hegel's language becomes much more severe—then there are no external alternatives and no non-existent possibilities because there is no other who could have been but is not. From the disposition of absolute necessity, there are no further ways that things could have been, no choice of alternatives, no outstanding contingencies that might or might not come about. This, however, is not because absolute necessity excludes possibility, as if only certain possibilities can manifest themselves while others cannot. On the contrary, the reason why there are no further possibilities is because that which is absolutely necessary already contains every possibility whatsoever. In this second sense of the term, absolute necessity is the least restrictive of the necessities because it is the ground for the actualization of the possible itself. What is absolutely necessary cannot be otherwise, not because there exists an other that it must not be, but rather because the actuality of this necessity already includes both the positive and the negative moments of possibility. This necessity is the expression of the possible as a universal, not only one side or another, with the other side outstanding, but all sides together from one disposition.

If something is absolutely necessary, the reason why it cannot be otherwise is because it already includes every possibility as the very constitution of what it is. This is why Hegel says that "[absolute necessity] is, therefore, because it is" (WL 215). In this version of modality, it is no longer only the case that each thing is because others are. Now Hegel says that each is because it is. Because of absolute necessity, being refers only to itself, and this leads to all of the following: (1) possibilities that are not immediately actual nevertheless exist as absolutely necessary, (2) the coexisting of unactualized possibilities becomes an affirmative field of multiplicity, and (3) being must perpetually move across itself.

27. Because absolute necessity is the total inclusion of every possibility whatsoever, absolute contingency is the final consequence of Hegel's argument.

> Sie ist daher es selbst, welche sich als Zufälligkeit bestimmt; - in ihrem Seyn sich von sich abstößt, in diesem Abstossen selbst nur in sich zurürckgekehrt ist, und in dieser Rükkehr als ihrem Seyn sich von sich selbst abgestossen hat. (WL 214)

> What this means is that necessity itself determines itself to be contingency: in its very being it repels itself from itself; in this very repelling it merely returns to itself; and in this return to being it has repelled itself from itself.

Necessity now determines itself as contingency. This is an unusual development from the necessity of premise twenty-four. At first, although absolute actuality contains every possibility as the embodiment of itself in actuality, this necessity only arises by excluding contingencies that fall outside of the set. If we think of this actuality as genus-actuality, then the actuality includes all possibilities of the genus, while at the same time excludes as mere contingency all possibilities that fall outside of this (either because they belong to another genus, or because they embody the extreme deficiencies and excesses beyond the determinate limitations of the set).[8] From the standpoint of premises twenty-three and twenty-four, absolute actuality is necessity because it determines itself as all possibilities within the set; however, this necessity does not yet determine itself as contingency because contingencies are precisely that which do not belong to the actuality. Although the actuality of a horse embodies every possibility of being a horse, the possibility of being a frog or a goat is from the disposition of this individual substance a mere contingency since these possibilities extend beyond what a horse is and can do.

Once Hegel introduces absolute possibility at premise twenty-five, thought can no longer distinguish the limited totality of possibilities that exist within any given set from the seemingly unruly contingencies that exist beyond the set. An actuality of absolute possibility recognizes every contingency as the possibility of being. Although possibilities are still defined by their ends in actuality, if this actuality is being itself, rather than one or another specificity of being, then possibilities go in every direction because they are no different than contingencies. While this conclusion might seem quite problematic, Hegel can be interpreted to give a reasonable explanation for how to think the unity of necessity and contingency in his two-sided consequence of affirmative multiplicity and absolute conversion.

> Die absolute Nothwendigkeit ist so die Reflexion oder Form des Absoluten; Einheit des Seyns und Wesens, einfache Unmittelbarkeit, welche absolute Negativität ist. Einerseits sind ihre Unterschiede daher nicht als Reflexionsbestimmungen, sondern als seyende Mannichfaltigkeit, als unterschiedene Wirklichkeit, welche die Gestalt von selbstständigen Anderen gegen einander hat. Andererseits da ihre Beziehung die absolute Identität ist, ist sie das absolute Umkehren ihrer Wirklichkeit in ihre Möglichkeit und ihrer Möglichkeit in Wirklichkeit. (WL 215)

> Absolute necessity is thus the reflection or form of the absolute, the unity of being and essence, simple immediacy which is absolute negativity. On the one hand, its differences do not have reflection as their determination, but are

> determined, instead, by affirmative multiplicity, a compartmentalized actuality, which carries with it the self-subsistent shape of others, which in turn stand against each other. On the other hand, since its relation is the relation of absolute identity, it is the absolute conversion of its actuality into its possibility and of its possibility into its actuality.

That absolute necessity has taken over the role that possibility had played as the reflection of the actual into itself leads to a two-sided consequence. The first side of the consequence is affirmative multiplicity. Affirmative multiplicity (*seiende Mannichfaltigkeit*) is the field of actualizability. This field is full of compartmentalized actualities (*unterschiedene Wirklichkeit*) whose own self-subsistent shape (*Gestalt*) is the shape of others. The second side of this consequence is absolute conversion. By absolute conversion (*absolute Umkehren*), Hegel explicitly acknowledges the complete actualization of possibility.

Affirmative multiplicity is a development from the limitless multiplicity of formal modality, as from the limited multiplicity of real modality, because it is both the limitation of existing in one way or another and the full presence of essence, of all possibilities together. On the one hand, affirmative multiplicity is only a multiplicity if it contains formal and real differences within it. A is still not -A. A is still not B. If these distinctions were no longer in place, multiplicity would be just gray unity without diversity and with no compartments. Then nothing would be possible because everything would be one. But on the other hand, this multiplicity is affirmative because the differences that reside within this multiplicity no longer stand in opposition against others, nor can find the positing of themselves to be the contradiction of any other one.

Each differentiated actuality, if it follows the logic of affirmative multiplicity, is the absolute conversion of actuality and possibility. Take an individually existing organism, what Aristotle calls primary substance, a goat grazing in the pasture.[9] Of course, the individual goat grazing in the pasture cannot go in every possible direction, nor make all of her possibilities actual. If she is in the pasture, she cannot possibly be standing with the other goats in the barn. If she is a goat, she cannot in the same manner not be a goat. Hegel has already established these contradiction-limitations with his theories of formal and real modality. In this sense, actuality and possibility are not absolute conversion. Each concept causes the other to fall short of itself.

But if we think about what this individual goat is, we establish her existence and her essence as one. In terms of her essence, the individually existing goat can go in every possible direction (relative to what a goat can be). She can graze in the pasture and stand with the other goats in the barn. Her individual existence is like a metaphysical gateway that directly corresponds to all of her possibilities as her essence. As the existence of her essence, she can receive all sorts of contrary determinations. Since she is a goat, she is both always determinate (always on the hill, in the barn, at the pasture, always here or there) and yet free from the limitations of her determinateness (her being on the hill does not exclude the possibilities of contraries in her essence). She is free determinateness because her individual existence as this goat is at the same time the totality of her possibilities as her essence.

But by the absolute conversion of actuality and possibility, Hegel means even more than the metaphor of a gateway between existence and essence. In this way his theory is distinct from Aristotle's theory of first-order actuality as the many substances. Hegel literally means that existence and essence are the same. What is this individually existing goat? She is "a" goat. The indefinite article exposes the individual to the universality of possibility. It is true that without existence, essence would be empty of determinateness, and that without essence, existence would have no movement or possibility. But Hegel means absolute conversion. The modal concepts are the same, not just mutually dependent.

In the subsequent section of the *Logic*, the "Relation of Substantiality," Hegel will call this relation of absolute conversion between actuality and possibility "actuosity" (*Actuosität*) (WL 220). Actuosity is a development upon the absolute conversion of actuality and possibility because it emphasizes not only the self-movement of substance, but also the tranquility of this movement. Being is substance that is differentiated from itself; however, it is differentiated only as itself; it moves as if against an other, but this movement is the quiet, unresisting movement of itself across itself. Yet, this quiet self-movement is nevertheless a movement that happens across affirmative differences. The differences in the substantiality of being are all the more pronounced, and the movement is all the more agitated, if in being other than itself, we recognize that each thing is the same being all the way through.

The absolute conversion of actuality and possibility therefore causes the most intense agitation of movement in every possible direction. Since the individual goat is its genus-being, not just a member of its genus, but the living existence of essence, the individual goat must receive all contraries as the totality of its possibilities. But only the logic of affirmative multiplicity could allow for the reception of all contraries; otherwise, the difference of each contrary would cause contradiction. Goats move about because existence and essence are one. They grow and pass away because they must expose what they are. They are sometimes standing, sometimes sleeping, sometimes eating, sometimes playing. Goats must exhibit all of these possibilities to be what they are. If existence and essence are one, then each instance of existence is the totality of essence. It is, as Leibniz claims, "all of one piece, like an ocean."[10]

Existence and essence are the same only if all contingency has been removed from affirmative multiplicity. If any contingency were to remain, then the field would not produce free movement because the remainder of contingency would dictate the determinate order of things. This is why Hegel says that this field is absolute necessity. Necessity is the essence of the free play of differentiated actualities in the field of affirmative multiplicity. Hegel writes,

> Die Nothwendigkeit als Wesen ist in diesem Seyn verschlossen; die Berührung dieser Wirklichkeiten durch einander erscheint daher als ein leere Aeusserlichkeit; die Wirklichkeit des einen in dem andern ist die nur Möglichkeit, die Zufälligkeit. Denn das Seyn ist gesetzt als absolut nothwendig, als die Vermittlung-mit-sich, welche absolute Negation der Vermittlung-durch-anderes ist, oder als Seyn das nur mit dem Seyn identisch ist; ein Anderes, das im Seyn Wirklichkeit hat, ist daher als schlechthin nur Mötliches, leeres Gestztseyn bestimmt. (WL 216)

> Necessity as essence is concealed in being. The contact that these actualities have with each other appears to be like an empty externality. The actuality of one in the other is only possibility, contingency. For being is posited as absolutely necessary, as the self-mediation which is the absolute negation of mediation-through-another, or being which is identical only with being. An other that has actuality in its being is, therefore, only possibility as such, determined to be empty positedness.

At the same time, however, the field only removes all contingency if it includes all contingency whatsoever. Affirmative multiplicity is thus the most powerful structure of contingency, in the sense that everything within this field both can be and is the other of itself.[11] By the free play of its determinateness, each actuality can give no reason but itself; its restrictions, self-imposed, come directly from being. Each actuality is only itself because each is the shape of all others. Although each actuality is inherently determinate, this determinateness is open to all contingencies because it is grounded only in being.

The field of affirmative multiplicity makes everything necessary, but this structure only works from the most fluid of all contingencies. Hegel thus concludes that contingency is absolute necessity. He says that if necessity is the essence of these free actualities, this is the case because necessity and contingency are the same. Free movement happens only because there is nowhere to move.

If Hegel were arguing simply from the disposition of things, then the claim that everything moves in every possible direction would have to face a number of intuitive objections. It would have to face the objection that if something cannot be otherwise, it should not then be free to go otherwise. And it would have to face the objection that a thing cannot move in every possible direction, but only in certain directions, while leaving beside itself various unactualized possibilities that never become realized. Although one might come to agree with Hegel's argument from formal and real modality, that unactualized possibilities exist, one might still remain skeptical of the more controversial claim, that each thing must be the actualization of possibility as one totality. However, the disposition that Hegel now requires is not simply one of abstract things in their self-reflexive identities, nor is it that of real things that are because others are. What saves Hegel from these objections is that he is arguing from the disposition of inclusive, absolute necessity, rather than from the disposition of things. He writes,

> Aber diese Zufälligkeit ist vielmehr die absolute Nothwendigkeit; sie ist das Wesen jener freyen, an sich nothwendigen Wirklichkeiten. Dieses Wesen ist das Lichtscheue, weil an diesen Wirklichkeiten kein Scheinen, kein Reflex ist, weil sie nur rein in sich gegründet, für sich gestaltet sind, sich nur sich selbst manifestiren,—weil sie nur Seyn sind.—Aber ihr Wesen wird an ihnen hervorbrechen und offenbaren, was es ist und was sie sind. Die Einfachheit ihres Seyns, ihres Beruhens auf sich, ist die absolute Negativität; sie ist die Freyheit ihrer scheinlosen Unmittelbarkeit. Dieses Negative bricht an ihnen hervor, weil das Seyn durch diß sein Wesen der Widerspruch mit sich selbst ist;—und zwar gegen diß Seyn in der Form des Seyns, also als die Negation jener Wirklichkeiten, welche absolut verschieden ist von ihrem Seyn, als ihr Nichts, als ein eben so freyes Andersseyn gegen sie, als die Negation jener Wirklichkeiten, welche absolut verschieden ist von ihrem Seyn, als ihr Nichts, als ein eben so freyes Andersseyn gegen sie, als ihr Seyn es ist. (WL 216)

> But this contingency is absolute necessity. It is the essence of those free, necessary actualities. This essence is light-shy, because in these actualities nothing can shine through, because there is no reflection, because they are grounded purely in themselves, shape themselves, manifest themselves only to themselves—because they are only being. But their essence will still break through and reveal what it is and what they are. The simplicity of their being, that they have recourse only to themselves, is absolute negativity; this is the freedom of reflectionless immediacy. Negativity breaks through in these actualities because being, since it is at the same time essence, is of its own accord contradiction. They break through against this being in the form of being, therefore at the same time as the negation of those actualities, a negation absolutely different from their being. They break through as their nothing, as an otherness which is just as free as their being is free.

If absolute necessity is the necessity of all possibilities, this type of necessity is then nothing other than contingency. As the complete unity of actuality and possibility, being must go in every possible way. This is indeed the final stand. By absolute necessity, if being goes in every possible way, there is nowhere further left to go. All possibilities become necessary. But Hegel then finds the strange but exciting result that the ability to-be-otherwise is secured only from the complete foreclosure of otherness. The structure of absolute necessity, because there is no further otherness or contingency, yields the most fluid openness of movement, an absolute necessity that is just as certainly absolute contingency, a whole field of differences and determinations whose narrowness has become so complete that the restrictions this necessity had imposed upon actualization are no longer restrictive.

It is therefore necessity that generates the freedom to-be-otherwise. Each actuality is free to be other than itself because in the process of being other than itself, each is itself. More than this, these actualities are only what they are if they are also the totality of all others. Necessity is the force that compels the freedom of their movement. Each actuality must be the freedom of itself in all others. This is the imperative of their freedom. They are completely taken by this imperative. They have no chance to opt out or find another way. From the disposition of absolute necessity, each actuality can only be itself insofar as it is simultaneously the totality of everything whatsoever. If each is the totality of everything, then no matter what each becomes in terms of determinate content, each cannot be otherwise than this that it is. Yet, precisely because each is the totality, each is also free from this determinateness.

When we recognize the goat as *a* goat, we appeal to her essence, but we really appeal to her contingency. To be what she is, she must be other than this immediate facticity, other than this individual goat standing at the fence. But this being otherwise is at the same time the original position of her immediate facticity as an individual. When we recognize her as a goat, we let the contingency "shine through" her. But what shines through the individual goat is more than her genus as an animal. Being itself shines through. This goat is. A goat cannot be other than this. But at the same time this necessity is the goat's absolute freedom to be otherwise. Contingency shines through at the level of being. Being makes the goat inherently necessary and completely free to be otherwise. This is what Hegel means by free actuality. The determinateness of being this is at the same time free otherness. All determinate things perish at the sign of this essence. The goat attempts to sustain herself against the sheer magnitude of her being. She follows the narrow path of her conditions. But just as she feels the force of her conditions, commanding her to eat and sleep and care for her own, she also feels the force of her ultimate necessity in being, and this is the force of her own destruction in contingency. This is why Hegel says that these actualities "perish" as they go free:

> Dieser ist das Maal, das die Nothwendigkeit, indem sie, welche absolute Rückkehr in sich selbst in ihrer Bestimmung ist, dieselben frey als absolut wirkliche entließ,—ihnen aufdrückte, worauf sie als den Zeugen ihres Rechts sich beruft, und an dem sie ergriffen nun untergehen. Diese Manifestation dessen, was die Bestimmtheit in Wahrheit ist, negative Beziehung auf sich selbst, ist blinder Untergang im Andersseyn; das hervorbrechende Scheinen oder die Reflexion ist an den Seyenden als Werden oder Uebergehen des Seyns in Nichts. Aber das Seyn ist umgekehrt eben so sehr Wesen, und das

> Werden ist Reflexion oder Scheinen. So ist die Aeusserlichkeit ihre innerlichkeit, ihre Beziehung ist absolute Identität; und das Uebergehen des Wirklichen in Mögliches, des Seyns in Nichts ein Zusammengehen mit sich selbst; die Zufälligkeit ist absolute Nothwendigkeit; sie selbst ist das Voraussetzen jener ersten absoluten Wirklichkeiten. (WL 216–17)

> This content is the mark that necessity impresses upon these actualities when it lets them go free as absolutely actual—for its determination is to be an absolute return into itself. This is the mark that necessity appeals to as witness to its right, and smitten by it, the actualities now perish. This manifestation, which is what the determination is in truth, is a negative relation to itself, and as such, is a blind destruction in otherness. The shining or reflection that breaks through is a becoming, in other words, a transition of being into nothing. But being is, to the contrary, at the same time essence, and becoming is at the same time reflection or a shining. Therefore, the externality is its inwardness; its relation is one of absolute identity, and the transition of actuality into possibility, of being into nothing, is a going-together of itself with itself. Contingency is absolute necessity. It is itself the presupposing of that first, absolute actuality.

The metaphors in this last passage of the chapter are suggestive but difficult to interpret. Hegel says that necessity leaves a "mark" (*Maal*) like a "witness to its right" (*Zeugen ihres Rechts*) and that these free actualities are "smitten" (*ergriffen*) at the sign of this mark and thereby "perish" (*untergehen*). On my reading, the point of these metaphors is to extend Hegel's analysis of two separate dispositions. One is the disposition of things, a disposition that Hegel criticizes throughout the chapter, where if something is identical with itself, it cannot also be the opposite of itself. But the other is the disposition of affirmative multiplicity, Hegel's primary conclusion and main focus in terms of "absolute necessity," a disposition where the law of non-contradiction no longer has an influence because we have come to include all possibilities from one standpoint.

From the disposition of things, then, there is no good reason why things have the content that they have. This content appears upon them like a mark appears upon a body. They simply are *because* they are. When I look at a map of the earth, I see the mountains and lakes as already there. Certainly I can attempt to explain how they came to be from the earth's formation at some prior point in time. But these reasons are always only relative. Ultimately, I can only say that the mountains and the lakes are because they are. This appeal to self-relation does not, however, diminish the force of their necessity. What it does, instead, is place the reason for things beyond the disposition of things. I can only say that although what immediately appears as the fact of existence is given by the force of necessity, the determinateness of this content nevertheless comes from beyond the internal logic of the content, as if from nowhere, like a mark left on a body.

If the absolute necessity in the determinate content of things comes from the relation of being alone, then each thing is bound to necessity but is also free to become otherwise because what makes each thing necessary is being alone. The bodies of these actualities carry on them the history of prior necessity, but this necessity no longer restricts them from being both themselves and all others. This radical access to possibility—where each thing is an aspect of everything, where each actuality affirms possibility in its entirety—is at the same time the mark of destruction in sheer otherness.[12] The mark allows them to disassociate themselves from the facticity of being only one or the other of what they are, and in this freedom, to be possibility *qua* possibility, but by this same mark, they have committed themselves as things to the unsustainable regions of sheer possibility.

Hegel concludes from this that from the disposition of things, being is self-contradictory. Contingency "breaks forth" (*bricht an ihnen hervor*) from the determinateness of all things because contradiction is the nature of being. Just as genus-essence receives all contraries of individual things, being receives not only the contraries of the genus, but the contraries of everything whatsoever. Being commits each thing to the abyss of its freedom in all other things. This is expressed foremost in the utter destruction of determinateness and in the transgression of every distinct actualization of itself. This radical possibility is at the same time access to the dark content of absolute negativity. The only stability things have is in the utter contingency of being itself.

CONCLUSION: ABSOLUTE NECESSITY AND CONTINGENCY ARE THE SAME.

Hegel's conclusion at premise twenty-seven that necessity and contingency are the same might seem too strong to be true. The dialectic between necessity and contingency that results from the formal and real premises of his argument presents us with the absolute conversion of actuality and possibility, where everything is necessary because everything is free to be otherwise. If modality leads to the utter instability of determinate being as the full presence of possibilities existing in actuality, this is no doubt an indication that modality does not expose the ultimate truth about ontology. We can then view this extreme result, where everything is necessary because everything is otherwise, as the flaw of modality, and as the main reason why substantiality, causality, and reciprocity emerge as the next concepts of the *Logic*. These concepts explain more about the dialectical role of stability in the interactions of affirmative multiplicity than modality can explain. However, by showing how substance comes out of modality in the first place, Hegel has also solved the problem that he had stated at the outset of the "Actuality" chapter, of how to recognize the ontological status of possibility as constitutive of reality. This is one of the primary, self-contained results of the chapter, and can be taken to be quite significant in its own right.

Hegel's conclusion is not only that possibilities necessarily exist as part of substance. His conclusion is also that because of the movement inherent to inclusive necessity, unactualized possibility is no longer a static category because actuality and possibility have become one unity. This leads to the consequence that contingency is inherently necessary to Hegel's system, in the sense that free movement can only happen from within a structure where nothing can be otherwise because everything is already in play. But this also leads to the consequence that thought can express the totality of possibility (both the positive and the negative side) in one actuality.

The traditional reading of Hegel's dialectic between absolute necessity and contingency is that Hegel ultimately recognizes the concept of contingency as a necessary element of his system. This is the most popular interpretation of Hegel's conclusion about modality, since it leads to a reassessment of the role of contingency in Hegelian thinking. Contingency turns out to be one of many necessary concepts. It is of absolute necessity in the sense that it must be included alongside the hundreds of other concepts that Hegel necessarily deduces from the presuppositionless opening of the *Logic*. This popular reading proposes that the system itself cannot be otherwise, that the concepts of the *Logic* are determined rationally by necessity, not by contingency, but that, in a paradoxical way, there also exists a place for the concept of contingency among the other concepts. This is paradoxical because, if contingency is one among many necessary concepts, the question then arises of whether the presence of contingency opens the way for new alterations of the otherwise seemingly rigid deduction of the concepts.

However, the close textual analysis I have outlined in this book reveals another viable conclusion about the necessity of contingency. By absolute necessity, Hegel proposes that the reason why things cannot be otherwise is because everything has already been included to the extreme point of absolute contingency. This conclusion is apparent from the inclusive nature of contradiction, a modal version of which Hegel prioritizes in the "Actuality" chapter. There is no further position above and beyond the contradictory position of *A is -A*. To include the negative along with the positive side is of absolute necessity because *A is -A* anticipates every possible permutation in every which way of A. This type of necessity comes from the inclusion of the negation, rather than from some determinate limit, which would restrict what can and can not be. Absolute necessity closes off the possible field of determinations, not by limiting this field to some determinate content at the expense of others, but rather by maintaining the utter and complete openness of negation. Absolute necessity is, therefore, contingency. The inclusion of the negative makes things unable to be otherwise because they are in every which way already otherwise than they are.

My analysis of Hegel's modal argument attempts to defend this alternative reading of the dialectic between absolute necessity and contingency and to show that this interpretation is as viable as the traditional reading that contingency is one of many necessary concepts in Hegel's system. While there is certainly textual evidence to support the traditional reading of Hegel's conclusion, and while this reading does lead to the exciting revision of Hegel as a thinker who includes contingency in his work, there are also issues with this traditional reading. If Hegel means that contingency is only one of many necessary concepts, necessity turns out to be more primary than contingency, in the sense that all concepts are of necessity and come from necessary developments. In contrast to necessity, contingency plays only a marginal role as merely one of these necessary concepts. But the reading I propose is more robustly dialectical. Absolute necessity turns out to be, in the most genuine sense, absolute contingency. This reading is also more obviously applicable to Hegel's conception of substance, which follows after the "Actuality" chapter as the culmination of the *Doctrine of Essence*. This reading is also much more applicable to the thematic of the *Doctrine of Essence* generally because it explains the modality behind how something is both universal and individual at the same time.

NOTES

1. Other non-organic examples work just as well—such as the moral disposition of a person's character, the requirements of a legal solution to a property infringement, the reason why Facebook has status updates, and so on. I have chosen to draw from the example of the plant and the seed because this is a process that Hegel was particularly fond of invoking. For Hegel's extensive treatment of "Plant Nature," see §343–49 of G. W. F. Hegel, *Philosophy of Nature*, translated by A.V. Miller (Oxford: Oxford University Press, 1970), 303–51. For an excellent study of contradiction and organisms, which offers an analysis of Hegel's treatment of plants, see Songsuk Susan Hahn, *Contradiction in Motion: Hegel's Organic Concept of Life and Value* (Ithaca: Cornell University Press, 2007). Also see Johann Wolfgang von Goethe, *The Metamorphosis of Plants* (Cambridge: MIT Press, 2009).

2. Leibniz, *Philosophical Essays*, 221.

3. Hegel's discussion of genus as an organic concept appears later in the *Logic*, in the "Life" chapter of the *Doctrine of the Concept* (SL 179–91).

4. For Levinas's definitions of ontology and metaphysics, see Emmanuel Levinas, *Totality and Infinity*, translated by Alphonso Lingis (Pittsburgh: Duquesne University Press, 1969), 42–48.

5. While MacDonald offers a persuasive account of how to approach the possible *qua* the possible, he attributes the sort of analysis one might give of absolute possibility to Adorno rather than to Hegel, while rejecting Hegel for being a philosopher who cannot think possibility. But to give this reading of Hegel, MacDonald has had to skip over the absolute possibility step of Hegel's argument. See, MacDonald, "Adorno's Modal Utopianism."

6. In the first introduction to *Being and Time*, Heidegger criticizes Hegel's concept of Being as it appears initially in the *Logic*. He objects that Hegel has not properly exposed the question of Being, but rather begins from a presumptive answer to this question. Of the three interpretations that Heidegger lists in the passage, he attributes the connotation of universality to Hegel (the other two interpretations are "Being as indefinable" and "Being as self-evident.") Heidegger, *Being and Time*, 22–23.

7. Houlgate discusses this same point but in terms of the being of nothingness. "Even if there were nothing," he writes, "there would still be being, since nothingness would still be what there is" (NC 46). He argues that for Hegel, being itself cannot be contingent. If there is nothing, then equally there is being. In contrast to Houlgate's analysis, Burbidge claims that when Hegel introduces absolute possibility, he introduces the question of whether there could have been nothing instead of something (SC 88).

8. Di Giovanni calls this the contingency of classical metaphysics. "In classical metaphysics, contingency was thought to be the result of a discrepancy between possibility and actuality" (CC 186). One of Di Giovanni's main contributions to the literature is to establish Hegel's theory of contingency as a critique of the classical. Di Giovanni argues that, for Hegel, reality would not be self-sufficient if it did not contain all irrationalities (that is, contingencies) as part of its own constitution (CC 197). This means that contingency is not derivative of being, but is rather necessary for the self-relation of being. I am quite sympathetic of Di Giovanni's reading in this respect.

9. For Aristotle's discussion of primary and secondary substances, see Chapter 5 of the "Categories" (CWA 2a13–4b20).

10. Leibniz, *Theodicy*, 131.

11. Lampert outlines this result when he claims that absolute necessity is "the interaction of all forces in every possible way—such that what *is* must continue to generate and envelop ever differing possibilities" (FM 75).

12. In his discussion of absolute necessity as that which makes all finite things perish, Houlgate invokes an infamous passage from Hegel's *Introduction to the Philosophy of History:* "There is nothing that history can be *except* 'the slaughter-bench on which the happiness of peoples, the wisdom of states and the virtue of individuals have been sacrificed'"(NC 48).

Appendix

A Premise-by-Premise Overview of the "Actuality" Chapter

As a further tool for analysis, I offer this premise-by-premise overview as a way to outline Hegel's basic argument in the chapter. Although dense, this version is designed to give the reader a detailed table of contents of the premises and to clarify the general trajectory as the dialectical inferences unfold step by step.

1. What is actual is existence.

    a. Premise one is self-evident. Look to existence to recognize what is actual.
    b. Existence means both the immediate fact of existence but also the totality out of which something emerges.

2. What is actual is possible.

    a. Premise two is also self-evident. If something is actual, then of course it is possible that it be actual.
    b. But this also means that the actual is identical with the possible in the sense that neither mode adds anything to the content of things.
    c. Hegel's entire argument refers back to these two self-evident premises. All subsequent premises can be viewed as equivalences, inferences, and conclusions built upon these two premises.

3. If what is actual is possible, then the possible is the reflection of the actual into itself.

    a. Premise three shows that premise two is the affirmation of premise one.
    b. Possibility is truth-affirming in the sense that if something is actual, then it is possible, which affirms its existence.
    c. If possibility reflects actuality into itself, then the actual is not only immediate, but is also mediated by possibility.
    d. But this mediation is nothing other than the actual's own identity-into-self.

    The sub-argument from possibility as necessary but not sufficient:
    i. If what is actual is possible, then what is not possible is not actual.
    ii. If what is not possible is not actual, then possibility is the most basic and necessary condition for any actualization whatsoever.
    iii. However, because possibility is only the affirmation of actuality into itself, what is actual is necessarily possible but what is possible is not necessarily actual.
    iv. Therefore, possibility is a necessary but not a sufficient condition for actuality.

4. But if the possible is only the reflection of the actual, it lacks itself and requires its completion in actuality.

    a. As long as thought fixates upon the actual, the possible is only the affirmation of the actual into itself and nothing more (equivalence of P3).
    b. However, since the possible affirms the actual, thought can also fixate upon the possible and not only upon the actual.
    c. When thought fixates upon the possible, the possible is still the identity of the actual, but of the actual *when it is not itself.*
    d. Therefore, because the possible affirms the actual in existence, it is also the actual when it lacks itself.
    e. Therefore, what is actual is complete and what is possible requires its completion in actuality.

5. Because it is reflection, possibility is the relating ground between the actual and the negation of the actual.

a. If A is possible, then A is A (equivalence of P3).
b. But A is A if and only if -A is -A (equivalence of P4).
c. If A is possible, then both A is A and -A is -A.
d. Therefore, possibility is the comparing relation between the A and the -A.
e. Therefore, the possible A contains the possible -A.
f. Therefore, the possible is the totality of the actual.
g. Problem: If what is actual is possible (P2), and what is possible contains the opposite of the actual (equivalence of P5), then does the actual also contain the opposite of the actual?

6. If the possible A contains the possible -A, then what is possible is also impossible.

a. If A is possible, then A can or can not be (equivalence of P5).
b. This means that if A is possible, then not A is also possible.
c. Therefore, because the possible is both the actual and the opposite of the actual, everything is possible if it is identical with itself.
d. But this also means that if something is not identical with itself, then it is impossible.
e. Therefore, everything is possible that does not contradict itself.
f. However, because what is possible is both the actual and the opposite of the actual, possibility contains diversity.
g. Since this diversity is both the actual but also the opposite of the actual, the diversity contains opposition.
h. Since what is actual is also possible (P2), but what is possible contains diversity, to actualize the diversity of the possible would turn the opposition into contradiction.
i. Therefore, everything together is self-contradictory and impossible.

7. Actualization cannot maintain this contradiction of the possible as the impossible.

a. This conclusion comes directly from the self-evidence of premises one and two.

i. What is actual is existence (P1).
ii. What is actual is possible (P2).

iii. But what is possible contains both the actual and the opposite of the actual (equivalence of P5).
iv. It would be a contradiction if something were to exist both as itself and as the opposite of itself.
v. Actuality is possibility, and possibility is both the actual and the opposite of the actual.
vi. Therefore, to actualize possibility is to actualize contradiction.

8. Because of this contradiction, actuality becomes reflected actuality. Reflected actuality is an actuality of possibility itself.

    a. To actualize the possible as a totality would mean to actualize contradiction (equivalence P7).
    b. However, to actualize only one strand or another of the possible, either the positive or the negative alone, would reduce the actual to a mere moment of the totality.
    c. But since reflected actuality is an actuality of the possible itself, it is neither self-contradictory nor only a moment of the totality.

9. The problem with reflected actuality is that since it is the possible itself, it is an actual that cannot become actual.

    a. Reflected actuality is an actuality of the possible itself (P8).
    b. However, the possible can only become complete in the actual (equivalence of P4).
    c. If reflected actuality is the possible itself, then it lacks itself and requires its completion in actuality.
    d. But this means that reflected actuality cannot become actual.

10. Hegel turns instead to contingent actuality, where the actual as what is immediately given posits its other as what could have been.

    a. It would be a contradiction to actualize the totality of possibility (equivalence of P7).
    b. However, to actualize only one strand or another of the possible would reduce the actual to an instance of existence, rather than to its totality.
    c. However, to let the actual take over the function of reflection as if it itself were the possible would make the actual incomplete and indeterminate (equivalence of P9).

d. But since contingent actuality is an immediate existent that posits the opposite of itself along with itself, it is not self-contradictory, nor is it merely a strand of the totality, but it is also not incomplete or indeterminate, as with reflected actuality.
e. Therefore, contingent actuality more effectively renders the possible in existence.

11. If the other of the actual equally exists, there is no reason why this actual is and why its other is not. Therefore, contingency has no ground.

a. An actual is contingent if it posits the equal existence of its opposite.
b. However, it would be a contradiction if something were to exist both as itself and as the opposite of itself.
c. Therefore, a contingent actuality maintains a relationship of indifference between its own existence and the existence of its opposite.
d. Therefore, a contingent actuality is an existent that is not it opposite, but that could have been its opposite if it were not itself.
e. Therefore, there is no reason why a contingent actuality exists and why its opposite does not exist.

12. But if the other of the actual equally exists, then actually depends upon what could have been.

a. If an actual is contingent, then the opposite of the actual could have been (equivalence of P11).
b. But this means that the actual is contingent only insofar as its opposite could have been.
c. Therefore, the contingency of the actual is dependent upon the equal existence of its opposite.
d. Therefore, contingency has ground.

13. Formal necessity is the source of these two arguments from contingency. Formal necessity is the coincidence of actuality and possibility.

a. If the actual depends upon the equal existence of its opposite (equivalence of P12), then there is not only an indifferent relationship between the existent actual and the existent opposite, but also one of coincidence.

b. Formal necessity is the coincidence of the possible as the actual.
c. This necessity is the only way to actualize possibility.

14. Real actuality results from the necessary form that actualization must take to actualize possibility.

    a. To actualize the totality of possibility in one actuality is to actualize contradiction (equivalence of P7).
    b. However, if formal necessity is the coincidence of the possible as the actual (P13), there is a way through mediation to actualize the diversity of possibility in one actuality.
    c. This mediation requires contextual opposition that can be overcome.
    d. Real actuality is the immediate existence of this contextual opposition.

15. Real actuality is real possibility.

    a. Real actuality is the immediate existence of contextual opposition (P14).
    b. But this means that the immediate existence can be overcome and that the actual can become the other of itself.
    c. Real possibility is both the reflection of this contextual opposition and the ability to overcome the opposition.

16. Real possibility is an existing multiplicity.

    a. Real actuality is an actuality of contextual opposition (equivalence of P14).
    b. Real actuality is real possibility (P15).
    c. Real possibility is the reflection of contextual opposition (inference of P14 and P15).
    d. If real possibility is the reflection of contextual opposition, then the actuality of this possibility generates existing bounded multiplicity.
    e. Therefore, real possibilities exist in the diversity of the contextual opposition.

17. Because of existing multiplicity, possibilities are dispersed in the actuality of others. The consequence is that something's possibilities are not its own but are always deferred to others.

a. This conclusion comes directly from the self-evidence of premises one and two.

i. What is actual is existence (P1).
ii. What is actual is possible (P2).
iii. But what is possible contains both the actual and the opposite of the actual (equivalence of P5).
iv. It would be a contradiction if something were to exist both as itself and as the opposite of itself (equivalence of P6).
v. Therefore, there is no formal way to actualize the totality of possibility in one actuality (equivalence of P7).
vi. Since real actuality also contains real possibility (equivalence of P15), and real possibility is the contextual opposition of this actuality, the only way to avoid the formal contradiction (at P7) is to disperse the possibilities in other actuals.
vii. Therefore, real possibilities exist dispersed in the actuality of others.
viii. This dispersion is the realization of existing multiplicity.

18. Since the possibilities of dispersed actuality do not seem to reside anywhere at all, something becomes actual through possibilities that are dispersed in its conditions.

a. If the possibilities of one thing exist in the actualities of others (equivalence of P17), then the existence of one's own possibilities is always deferred and never seems to reside in any actuality whatsoever.
b. But if the possibilities in one thing are the conditions for the actualization of oneself in others, then the possibilities are both dispersed in others but also exist as one's own possibilities.
c. Therefore, immediate actualities are the conditions of possibility for the actuality of themselves in and as others.
d. Therefore, conditions are both actualities and possibilities together.

The sub-argument from possibility completion:
i. It would be a contradiction if something were to exist both as itself and as the opposite of itself (equivalence of P7).
ii. To avoid this contradiction, existing multiplicity disperses possibility into other actuals (equivalence of P17).

iii. Since its possibilities are in other actuals, something emerges through possibilities that are in other actuals (equivalence of P18).
iv. Therefore, something is not only immediately actual, but becomes actual by completing the conditions of its possibilities.

19. What is initial can only become actual if it does not contradict the conditions that make it possible.

    a. Something initial can only emerge into actuality if it completes the conditions of its possibilities, which exist dispersed in other actuals.
    b. If something does not complete the conditions that make it possible, it cannot become actual.
    c. Therefore, it would be a contradiction if something where to emerge into actuality without completing the conditions of its possibilities.

20. But since each condition contains a multiplicity of other actuals, to become actual is to become in contradiction.

    a. However, since something only emerges into actuality through possibilities that at first exist in others, the actuality that emerges is equally the actuality of others (inference from P18).
    b. Since the actuality that emerges completes the possibilities of other actuals, to become actual through conditions is to become both itself and its others.
    c. It would be a contradiction if something were both itself and its others (equivalence of P7).
    d. Therefore, when something becomes actual through the conditions that make it possible, it becomes this in contradiction.

21. Something initial becomes actual if all of its conditions are present. Therefore, what is really possible can no longer be otherwise. This possibility is real necessity.

    a. Something can only emerge into actuality through possibilities that at first exist in other actuals (equivalence of P18).
    b. If something has emerged into actuality, then it must have completed the conditions of its possibility.

c. Therefore, the conditions of possibility are really necessary.

22. Contingency is nevertheless the reason behind why what is really possible can no longer be otherwise.

    a. Something initial only emerges into itself through possibilities that exist in others (equivalence of P18).
    b. If it cannot find itself in and as others, then something cannot be itself.
    c. Something must be able not to be itself in order to be itself.
    d. Therefore, real necessity depends upon contingency.

23. Absolute actuality is an actuality of the entire process once all of the conditions are present. It is the unity of contingency and necessity.

    a. The possibility of something exists at first in the actualities of others (equivalence of P17).
    b. However, if something completes the conditions that make it possible, it emerges as itself in and as these others.
    c. Therefore, the possibilities only seem to be dispersed in others.
    d. Absolute actuality is the existence of this self-relation.

24. Necessity, not possibility, is the reflection of absolute actuality into in-itself.

    a. Absolute actuality is the self-relation of possibilities that had seemed to be dispersed in other actuals (equivalence of P23).
    b. Because the possibilities in others are the self-relation of the actuality, all possibilities of the actuality have become necessary to the actuality.
    c. Therefore, necessity, not possibility, is the reflection of absolute actuality into itself (inference of P3 and P23).

25. Necessity is the in-itself because absolute actuality already includes all possibilities. This actuality is as much possibility as it is actuality. Hegel calls this absolute possibility.

    a. Absolute actuality is the relation of all possibilities as an actuality (equivalence of P24).
    b. But this means that it is an actuality whose content is only possibility.

c. Hegel calls this realization absolute possibility.

26. If absolute actuality is absolute possibility, the necessity of this actuality becomes absolute necessity.

    a. Absolute actuality contains only those possibilities that are generated from the context of the relation.
    b. However, because the possibility of this actuality is absolute possibility, absolute actuality necessarily contains all possibilities whatsoever.
    c. Absolute actuality is in this sense self-contradictory. It contains only those possibilities generated from the context, and yet it contains all possibilities whatsoever.
    d. Hegel calls the resolution of this contradiction absolute necessity.
    e. Absolute necessity is the necessity of total inclusion.

27. Because absolute necessity is the total inclusion of every possibility whatsoever, absolute contingency is the final consequence of Hegel's argument.

    a. Since absolute necessity includes all possibilities, it includes even those possibilities that exist beyond the context of each absolute actuality (equivalence of P26).
    b. This means that absolute necessity includes all contingencies whatsoever.

Therefore, absolute necessity and contingency are the same.

Bibliography

WORKS BY HEGEL

Hegel, G. W. F. *Elements of the Philosophy of Right*. Translated by Allen W. Wood. Cambridge: Cambridge University Press 1993.

———. *The Encyclopaedia Logic*. Translated by T. F. Geraets, W. A. Suchting, and H. S. Harris. Indianapolis: Hackett Publishing, 1991.

———. *Enzyklopädie der philosophischen Wissenschaften im Grundrisse, 1830, Erster Teil die Wissenschaft der Logik Mit den mündlichen Zusätzen*. Frankfurt am Main: Suhrkamp Verlag, 1970.

———. *Hegel's Science of Logic*. Translated by A. V. Miller. Amherst: Humanity Books, 1969.

———. *Introduction to the Philosophy of History*. Translated by Leo Rauch. Indianapolis: Hackett Publishing, 1988.

———. *Lectures on the History of Philosophy: Volume 3, Medieval and Modern Philosophy*. Translated by E. S. Haldane and Frances H. Simson. Lincoln: University of Nebraska Press, 1995.

———. *Lectures on Logic*. Translated by Clark Butler. Bloomington: Indiana University Press, 2008.

———. *Phenomenology of Spirit*. Translated by A.V. Miller. Oxford: Oxford University Press, 1977.

———. *Philosophy of Nature*. Translated by A.V. Miller. Oxford: Oxford University Press, 1970.

———. *The Science of Logic*. Translated by George di Giovanni. Cambridge: Cambridge University Press, 2010.

———. *Werke in zwanzig Bänden*. Frankfurt am Main: Suhrkamp Verlag, 1969.

WORKS ON HEGEL

Baptist, Gabriella. "Ways and Loci Modality. The Chapter 'Actuality' in the *Science of the Logic* between Its Absence in Jena and Its Disappearance in Berlin." In *Essays on Hegel's Logic*. Edited by George di Giovanni, 127–44. Albany: State University of New York Press, 1990.

Beiser, Frederick C. *Hegel*. London: Routledge, 2005.

Brown, Nahum. "Indeterminacy, Modality, Dialectics: Hegel on the Possibility *Not to Be*." In *The Significance of Indeterminacy*. Edited by Robert H. Scott and Gregory S. Moss, 104–23. New York: Routledge, 2018.
———. "Transcendent and Immanent Conceptions of Perfection in Leibniz and Hegel." In *Transcendence, Immanence, and Intercultural Philosophy*. Edited by Nahum Brown and William Franke, 183–207. New York: Palgrave Macmillan, 2016.
Burbidge, John W. *The Logic of Hegel's Logic*. Peterborough: Broadview Press, 2006.
———. *Hegel's Systematic Contingency*. New York: Palgrave Macmillan, 2007.
Carlson, David Gray. *A Commentary to Hegel's Science of Logic*. New York: Palgrave Macmillan, 2007.
De Boer, Karin. *On Hegel: The Sway of the Negative*. New York: Palgrave Macmillan, 2010.
Di Giovanni, George. "The Category of Contingency in the Hegelian Logic." In *Art and Logic in Hegel's Philosophy*. Edited by Warren E. Steinkraus, 179–200. Atlantic Highlands: Humanities Press, 1980.
Hahn, Songsuk Susan. *Contradiction in Motion: Hegel's Organic Concept of Life and Value*. Ithaca: Cornell University Press, 2007.
Harris, Errol E. *An Interpretation of the Logic of Hegel*. Lanham: University Press of America, 1983.
Harris, H. S. *Hegel: Phenomenology and System*. Indianapolis: Hackett Publishing, 1995.
Henrich, Dieter. *Hegel im Kontext*. Frankfurt: Suhrkamp, 1971.
Hoffmeyer, John F. *The Advent of Freedom: The Presence of the Future in Hegel's Logic*. Cranbury: Associated University Presses, 1994.
Houlgate, Stephen. "Necessity and Contingency in Hegel's *Science of Logic*." *Owl of Minerva* 27 (1995): 37–49.
———. *The Opening of Hegel's Logic: From Being to Infinity*. West Lafayette: Purdue University Press, 2006.
Lampert, Jay. "Hegel on Contingency, or, Fluidity and Multiplicity." *Bulletin of the Hegel Society of Great Britain* 51.2 (2005): 74–82.
Longuenesse, Béatrice. *Hegel's Critique of Metaphysics*. Translated by Nicole J. Simek. Cambridge: Cambridge University Press, 2007.
Macdonald, Iain. "Adorno's Modal Utopianism: Possibility and Actuality in Adorno and Hegel." *Adorno Studies* 1 (2017): 2–12.
Marcuse, Herbert. *Hegel's Ontology and the Theory of Historicity*. Translated by Seyla Benhabib. Cambridge: MIT Press, 1987.
McTaggart, John and Ellis. *A Commentary on Hegel's Logic*. New York: Russell & Russell 1910.
Mure, G. R. G. *A Study of Hegel's Logic*. Oxford: Oxford University Press, 1950.
Ng, Karen. "Hegel's Logic of Actuality." *Review of Metaphysics* 63 (2009): 139–72.
Padui, Raoni. "The Necessity of Contingency and the Powerlessness of Nature: Hegel's Two Senses of Contingency," *Idealistic Studies* 40.3 (2010): 243–55.
Pinkard, Terry. *Hegel's Phenomenology: The Sociality of Reason*. Cambridge: Cambridge University Press, 1994.
Pippin, Robert B. *Hegel's Idealism: The Satisfactions of Self-Consciousness*. Cambridge: Cambridge University Press, 1989.
Russon, John. *Reading Hegel's Phenomenology*. Bloomington: Indiana University Press, 2004.
———. *The Self and Its Body in Hegel's Phenomenology of Spirit*. Toronto: University of Toronto Press, 2001.
Stekeler-Weithofer, Pirmin. "Hegel on Reality as a Modal Notion." In *Hegel's Analytic Pragmatism*. Unpublished. Accessed online: http://www.sozphil.uni-leipzig.de/cm/philosophie/files/2012/11/StekelerHegelsAnalyticPragmatism.pdf.
———. "Scientific Truth as Augmented Reality: On the Contrast between 'Wirklichkeit' and 'Actuality,'" *De Gruyter* (2017): 83–100.
Taylor, Charles. *Hegel*. New York: Cambridge University Press, 1975.
Winfield, Richard Dien. *Hegel's Science of Logic: A Critical Rethinking in Thirty Lectures*. Lanham: Rowman & Littlefield, 2012.

Yamane, Tomoyuki. *Wirklichkeit: Interpretation eines Kapitals aus Hegels "Wissenschaft der Logik."* Frankfurt am Main: Peter Lang, 1983.
Yeomans, Christopher. *Freedom and Reflection: Hegel and the Logic of Agency*. Oxford: Oxford University Press, 2012.

OTHER WORKS

Agamben, Giorgio. *The Coming Community*. Translated by Michael Hardt. Minneapolis: University of Minnesota Press, 1993.
———. *Homo Sacer: Sovereign Power and Bare Life*. Translated by Daniel Heller-Roazen. Stanford: Stanford University Press, 1998.
———. *Nudities*. Translated by David Kishik and Stefan Pedatella. Stanford: Stanford University Press, 2011.
———. *Potentialities: Collected Essays in Philosophy*. Translated by Daniel Heller-Roazen. Stanford: Stanford University Press, 1999.
———. *The Time That Remains: A Commentary on The Letter to the Romans*. Translated by Patricia Dailey. Stanford: Stanford University Press, 2005.
Aristotle, *The Complete Works of Aristotle*, 2 volumes. Edited by Jonathan Barnes. Princeton: Princeton University Press, 1984.
Aquinas, Saint Thomas. *Commentary on the Metaphysics of Aristotle, Volume 1*. Translated by John Patrick Rowan. Washington: H. Regnery Company, 1961.
Brouwer, L. E. J. "Intuitionism and Formalism." *Bulletin of the American Mathematical Society* 37 (1999): 55–64.
Brown, Nahum. "Aristotle and Heidegger: Potentiality in Excess of Actuality." *Idealistic Studies* 46.2 (2017): 199–214.
———. "Indeterminacy, Modality, Dialectics: Hegel on the Possibility *Not to Be*." In *The Significance of Indeterminacy: Perspectives from Asian and Continental Philosophy*. Edited by Robert H. Scott and Gregory S. Moss, 104–23. New York: Routledge, 2018.
———. "The Modality of Sovereignty: Agamben and the Aporia of Primacy in Aristotle's *Metaphysics* Theta." *Mosaic* 46 (2013): 169–82.
Casey, Edward S. *The Fate of Place: A Philosophical History*. Berkeley: University of California Press, 1997.
———. *Getting Back into Place: Toward a Renewed Understanding of the Place-World*. Bloomington: Indiana University Press, 1993.
De la Durantaye, Leland. *Giorgio Agamben: A Critical Introduction*. Stanford: Stanford University Press, 2009.
De Saussure, Ferdinand. *Course in General Linguistics*. Translated by Wade Baskin. New York: McGraw-Hill Paperbacks, 1966.
Deleuze, Gilles. *The Fold: Leibniz and the Baroque*. Translated by Tom Conley. Minneapolis: University of Minnesota Press, 1993.
———. *The Logic of Sense*. Translated by Mark Lester. New York: Columbia University Press, 1990.
Derrida, Jacques. *Speech and Phenomena*. Translated by David B. Allison. Evanston: Northwestern University Press, 1973.
Fitting, Melvin and Richard L. Mendelsohn. *First-Order Modal Logic*. Dordrecht: Kluwer Academic Publishers, 1998.
Goethe, Johann Wolfgang von. *The Metamorphosis of Plants*. Cambridge: MIT Press, 2009.
Heidegger, Martin. *Aristotle's Metaphysics Θ 1–3: On the Essence and Actuality of Force*. Translated by Walter Brogan and Peter Warnek. Indianapolis: Indiana University Press, 1995.
———. *Being and Time*. Translated by John Macquarrie and Edward Robinson. New York: Harper & Row, 1962.
———. *Poetry, Language, Thought*. Translated by Albert Hofstadter. New York: Harper & Row, 1971.
Hobbes, Thomas. *Leviathan*. Indianapolis: Hackett Publishing, 1994.

Hughes, G. E, and Cresswell, M. J. *A New Introduction to Modal Logic.* London: Routledge, 1996.

Kant, Immanuel. *Critique of Pure Reason.* Translated by Paul Guyer and Allen W. Wood. Cambridge: Cambridge University Press, 1998.

Leibniz, Gottfried Wilhelm. *Philosophical Essays*. Translated by Daniel Garber and Roger Ariew. Indianapolis: Hackett Classics, 1989.

———. *Theodicy*. Translated by E.M. Huggard. BiblioBazaar, 2007.

Levinas, Emmanuel. *Totality and Infinity*. Translated by Alphonso Lingis. Pittsburgh: Duquesne University Press, 1969.

Meinong, Alexius. "The Theory of Objects." In *Realism and the Background of Phenomenology.* Edited by Roderick M. Chisholm. Michigan: Free Press, 1960.

Silverman, Hugh J. *Inscriptions*. Evanston: Northwestern University Press, 1997.

Spinoza, Baruch. *The Ethics.* Translated by Samuel Shirley. Indianapolis: Hackett, 1995.

Stang, Nicholas F. *Kant's Modal Metaphysics*. Oxford: Oxford University Press, 2016.

Whyte, Jessica. "I Would Prefer Not To: Bartleby and the Potentiality of the Law." *Law and Critique* 20 (2009): 309–24.

Index

About the Author

Nahum Brown is a research fellow in the philosophy department at Sun-yat Sen University. He is the editor, along with William Franke, of *Transcendence, Immanence, and Intercultural Philosophy* (2016) and, with J. Aaron Simmons, *Contemporary Debates in Negative Theology and Philosophy* (2017). He is also the author of numerous journal articles and book chapters, including "Aristotle and Heidegger: Potentiality in Excess of Actuality" (*Idealistic Studies*), "The Modality of Sovereignty: Agamben and the Aporia of Primacy in Aristotle's Metaphysics Theta" (*Mosaic*), "Transcendent and Immanent Conceptions of Perfection in Leibniz and Hegel," and "Is Hegel and Apophatic Thinker?"

www.ingramcontent.com/pod-product-compliance
Ingram Content Group UK Ltd.
Pitfield, Milton Keynes, MK11 3LW, UK
UKHW041122071026
15587UKWH00009B/81

9 781498 560566